SOCIAL MEDIA ENGINE

Steven Keltsch

ISBN: 978-0-578-86539-3 (Paperback)

Any references to historical events, real people, or real places are used fictitiously. Names, characters, and places are products of the author's imagination.

A special thanks goes to my Mentor:
Dennis Yu, CEO of BlitzMetrics, and his team, contributed to, Manuscript, Images, Design, and Content.
Steven Keltsch CEO of Outrageous Digital, Author, Editor.
Printed by Outrageous Digital in the United States of America.

Outrageous Digital
100 West Big Beaver Road
Suite 200
Troy, Michigan 48084
www.outrageousdigital.com

Table of Contents

I'd like to take this opportunity to welcome you. I'm Steven Keltsch CEO of Outrageous Digital. Our agency specializes in helping small to medium size businesses get the most out of digital marketing.

Most of us have had or will have some type of opportunity that can change our lives. What's most important, **is what you do about it.**

During the pandemic, **COVID-19 of 2020,** I spent all of my time learning from my mentor, **Dennis Yu.** I have no doubt, Dennis is a genius. However, like most geniuses, they're complicated.

What this book reflects is a digital marketing plan that's **proven with a Billion Dollar ad spend budgets** As you read this book you will see testimonials from Kenny Lauer Former VP Marketing, Golden State Warriors.

You'll learn about the **6 phases of Social Media Engine.** How to setup what we call **plumbing,** which is measuring your success using analytics. How best to establishing your **goals.** Learn how to use **content** to drive conversions. I will explain in a step by step strategy, how **targeting** the right audience who's most interested in what you are selling. After you have found your audience how to scale through **amplification.** And finally how to use **optimization, learning** what's working and what's not working.

Checkout our podcast **"Social Media Engine,"** wherever you get your podcasts.

Join my Podcast, **SOCIAL MEDIA ENGINE** as we discuss how to get the most from this book!

Every Tuesday at 2:00 PM EST

Social Media Engine Cheat Sheet

Steven Keltsch
Business Owner
Outrageous Digital

Cheat Sheet:
Outrageous Digital Marketing

PLUMBING › GOALS › CONTENT › TARGETING › AMPLIFICATION › OPTIMIZATION

1 PLUMBING

- ☐ Facebook Public Figure Page.
- ☐ Facebook Ad through Business Mgr.
- ☐ Social Accounts.
- ☐ Ultimate Life Website.

2 GOALS

- ☐ Define your mission (link to the goals page of your strategy assessment).
- ☐ Income Goal.
- ☐ Funnel (Group or Private Coaching

3 CONTENT

- ☐ Why video.
- ☐ Content Library.
- ☐ 3x3 Grid.
- ☐ Topic Wheel.

4 TARGETING

- ☐ Determine persona targeting (like to targeting assessment).

5 AMPLIFICATION

- ☐ Boost Why video to 1 saved persona
- ☐ Boost 3x3 Grid videos to at least 1 persona.
- ☐ Boost Topic Wheel content to 1 persona.

6 OPTIMIZATION

- ☐ Boost optimization: 4 Stages.

Why Social Media Engine?

Facebook amplifies what's working, as we've learned from the Warriors case study that Facebook released. Custom audiences = people-based marketing = remarketing = sequences across channels = word of mouth at scale. Custom audiences allow us to market cross-channel to amplify where fans are already engaging in the stadium, in the app, in our email sequences, on our website, via our Google channels, and anywhere we can place a tag.

We have grown these audiences and nurtured them through funnel stages of Awareness > Consideration

> Conversion. The ticketing and merchandise revenue we drive is a direct function of how well we have nurtured these engagement audiences - for example, organic efforts during the pre-season and boosted posts have already preformed well organically.

> **"Facebook ads increase the yield of our other marketing programs. Specifically, custom audiences yield at least a 1,000% ROAS and were as high as 4,000% during the playoffs. That means getting back $40 in revenue for every $1 spent. Lookalikes on top of custom audiences throw fuel on the fire."**

Kenny Lauer, *Former VP Marketing, Golden State Warriors*

Set-up, Strategy, Scale!

The 6 Phase of Social Media Engine

Social Media Engine is an Amplifier of what's already working.

Social media advertising is less about advertising and more about paid **Word of Mouth**.

Third-party authority mentions and reviews convert better because they generate implied endorsement.

My social advertising relies upon my content marketing engine to generate and collect authority.

Relevancy is the intersection of Content and Targeting, whether Google Quality Score or on Facebook.

The chaining of messaging is called Sequencing, also known as Marketing Automation in the email space and People-Based Marketing by Facebook.

My company's social strategy is the unique intersection of **Goals**, **Content**, and **Targeting**. They are channel and tool independent.

Conversely, social media tactics change constantly, can be delegated, and are repeatable. They are channel and tool dependent, driven by checklists.

The three components of successful social selling are **People**, **Process** and **Platform** (technology). Not interchangeable.

There are 6 phases to Social Media Engine

Maximize what's already working in your business to increase conversion rates on existing traffic and find more ideal customers. It's not for brand new products or businesses that don't have an existing funnel.

1 PLUMBING

In **Digital Plumbing**, you build your audiences and track results. With reliable analytics, you can determine where an additional ounce of effort or dollar in ad spend can make the most difference

2 GOALS

Your **Goals** are your metrics and your mission. Metrics are your numerically driven targets - cost per lead, ROI, revenue, traffic, and so forth. Your mission must be authentically defined in the WHY of your brand authentic statements that drive content that converts at each stage of the funnel. These values drive content sequences that effectively drive traffic and conversion.

3 CONTENT

Your **Content** is tied to these funnel metrics and audience segments - posts to boost to drive ticket/ merchandise sales, sponsored content to drive partner activations, videos to drive database growth, and so forth.

4 TARGETING

Target your content to multiple owned audiences (Facebook remarketing, Google remarketing, email remarketing, app remarketing, tie-ins with your CRM) and core audiences (lookalike audiences for each conversion type, media/ influencer targets related interest targets).

There are 6 phases to Social Media Engine

Getting **Digital Plumbing** in place is key to being able to create these audiences and create cross-channel campaigns (people who've been to your site but haven't bought, who are fans but you don't have an email, have bought last year but not this year, are in your email list, but haven't been to the site, and so forth).

Once we have established the triad of **Goals, Content,** and **Targeting,** we're ready to run ads. **Amplify** the most important pieces of content that will attract the most relevant people and drive engagement; intensify promotional efforts ti the engaged crowd for conversions; place brand content in the News Feed of influences to incept the media

Finally, in **Optimization,** we constantly and repeatedly iterate. Stay in the game. Use analytics to determine spend. Expand on working audiences, tweak bidding and creatives where necessary, re-allocate budgets and always measure your performance in terms of your content and targeting against your goals to define success.

Strategy Audit (GCT)

Defining Your Goals, **Content**, and **Targeting** is your foundation for digital marketing.
What problem would you like solved?

In one sentence, describe your business's mission. Identify the desired outcome and your customer segments.

Which channels do you see as critical to your marketing goals?
(Select all that apply)

Current monthly ad spend:

☐ $100 ☐ $500

☐ $2,500 ☐ $10,000

How many people are on your team?

☐ Just me
☐ Me and some freelancers
☐ Have marketing team.
☐ Enterprise

	Already doing well/ Would like to improve/ Not yet doing	Rank satisfaction (1-5) *(1 being unhappy and 5 being quite pleased)*
Facebook		5
Google Ads		5
SEO		5
LinkedIn		5
Twitter		5
Email		5
Local reviews		5
Apps		5

#ACC: Awareness, Consideration, Conversion

If you have defined leads or sales as one of your goals, what is your target cost per action or return on ad spend? (Use historical data and estimate of your break-even cost).

Cost per Lead:

Cost per Sale:

Target Return on Ad Spend (*ROAS*):

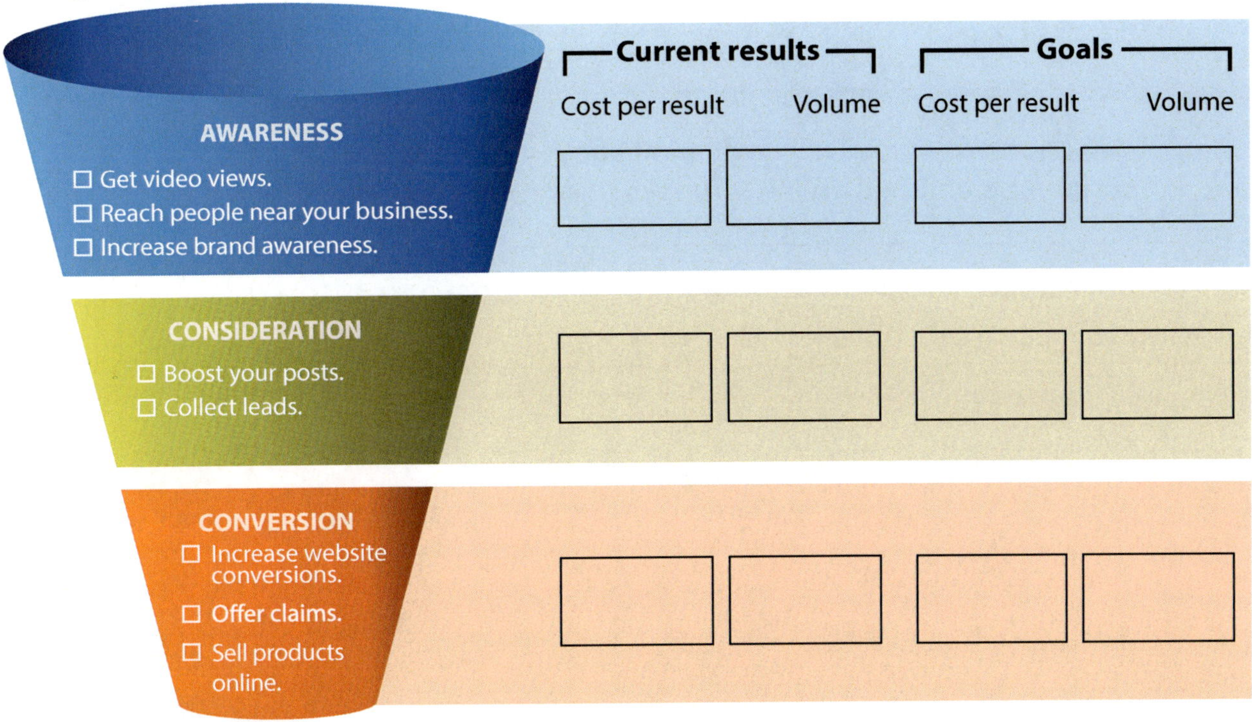

What is your story? (If your story is published online, like on your "about" page, write the link below.)

What is your business the BEST in? What niche do you aim to dominate?

Why do people love you? (If information about your Unique Selling Proposition is published online, write the link below.)

What credible proof do you have to tell your story? Is the proof online? Can you share it?

Demographics

Who is your customer? [gender, location, age group(s), profession(s), marital status, annual income, etc.]

Name three common interests your customers share.	1.
	2.
	3.
Name your three closest competitors.(Competitors your customers regard as a substitute, not necessarily the biggest competitors in your industry.)	1.
	2.
	3.
List 5 workplace-based targets. Think of the influences in your field, the jobs, titles of your customers, etc.)	1.
	2.
	3.
	4.
	5.

Your Marketing Technology Stack

Content Management System:
WordPress, Drupal, SiteCore, custom

none

Web Analytics:
Google Analytics, Analytics 360 Suite, Omniture (Adobe Analytics)

none

Tag Manager:
Google Tag Manager, Tealium, Adobe Tag Manager

none

Marketing Automation:
Infusionsoft, Marketo, HubSpot, Active Campaign, MailChimp, ExactTarget

none

Salesforce Automation:
Salesforce, Netsuite

none

Social Monitoring/Management:
Radian6, HootSuite, Sprinklr, Spredfast

none

Reporting:
Tableau, Domo, SimplyMeasured, SocialBakers

none

Ads Management:
Marin, WordStream

none

Landing Page Optimization:
Optimizely, Google Experiments, Leadpages

none

Shopping Cart:
Shopify, ZenCart, Yahoo! Cart, BigCommerce

none

ONE-PAGE CHECKLIST:

- [ ] Create your Facebook ads account using Business Manager.
- [ ] Create your Google Ads account.
- [ ] Create your Google Tag Manager (GTM) account.
- [ ] Set up your Google Analytics with GTM.
- [ ] Create your website custom audiences on Facebook.
  1, 7, and 28 day page and website engagement audiences.
- [ ] Set up Facebook pixel and standard events using GTM.
- [ ] Set up Google remarketing pixel using GTM.
- [ ] Set up Google Ads conversion tracking using GTM.
- [ ] Publish your GTM container and tags to your website.
- [ ] Verify that everything is set up properly.

Digital Plumbing Audit

Do you have your digital plumbing in place so you can track, measure, and re-target?

3 Content Production Strategies

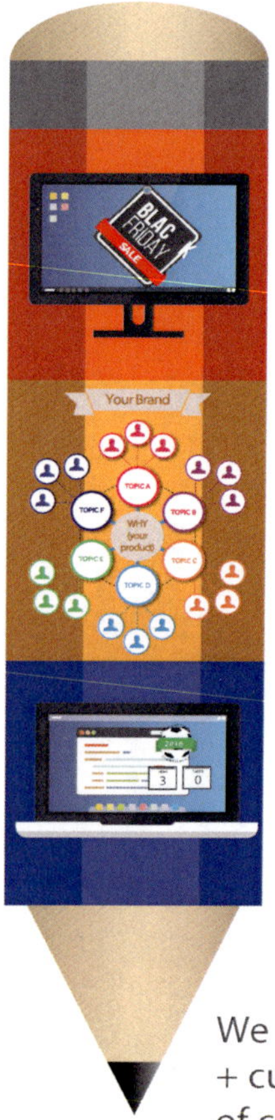

Content Calendar

A calendar is great for seasonal businesses and campaigns that are timed by date-- Black Friday, summer blowouts, product launches, and events. These are usually great revenue generators because they rely upon sales and stack upon the power of other channels (TV, radio, in-store, etc).

Topic Wheel

A Topic Wheel is fantastic for building evergreen content that are triggered by user action-- lead magnets, autoresponders, inbound marketing efforts, SaaS/recurring products, and software companies. This technique is excellent at building loyalty because it amplifies word of mouth into sales (collecting what customers are saying about you and distributing their words).

Spontaneous Content

Spontaneous content production is excellent for publishing businesses that have to produce hot, fresh content-- today's sports scores, commentary on current events, opinion pieces, and general blogging. This is the easiest place to start because it doesn't require a structure in advance, can be done by individuals ersus teams, and allows you the ¯exibility to jump on the latest trends.

We believe in a combo of a content calendar, the Topic Wheel, and spontaneous (news + curation) content working together. Most companies do just one of the three models of content production, but you'll have greater power when you combine all three.

Plumbing

- ✓ Create your Facebook Ads account using Business Manager
- ✓ Create your Google Ads account and tie it to the Google My Client Center (MCC).
- ✓ Create your Google Analytics account.
- ✓ Create your Google Tag Manager (GTM) account.
- ✓ Publish your GTM container and tags to website. Facebook Instant Articles.
- ✓ Google AMP (Accelerated Mobile Pages).

☐ **Fundamental** ☐ **Advanced**

We need to get your ads and analytics in sync-- the data and tracking audiences, that we're feeding all the right signals into Facebook/Google, and the flowing property as illustrated here. So, when the numbers match up, that's when building our ads are effective.

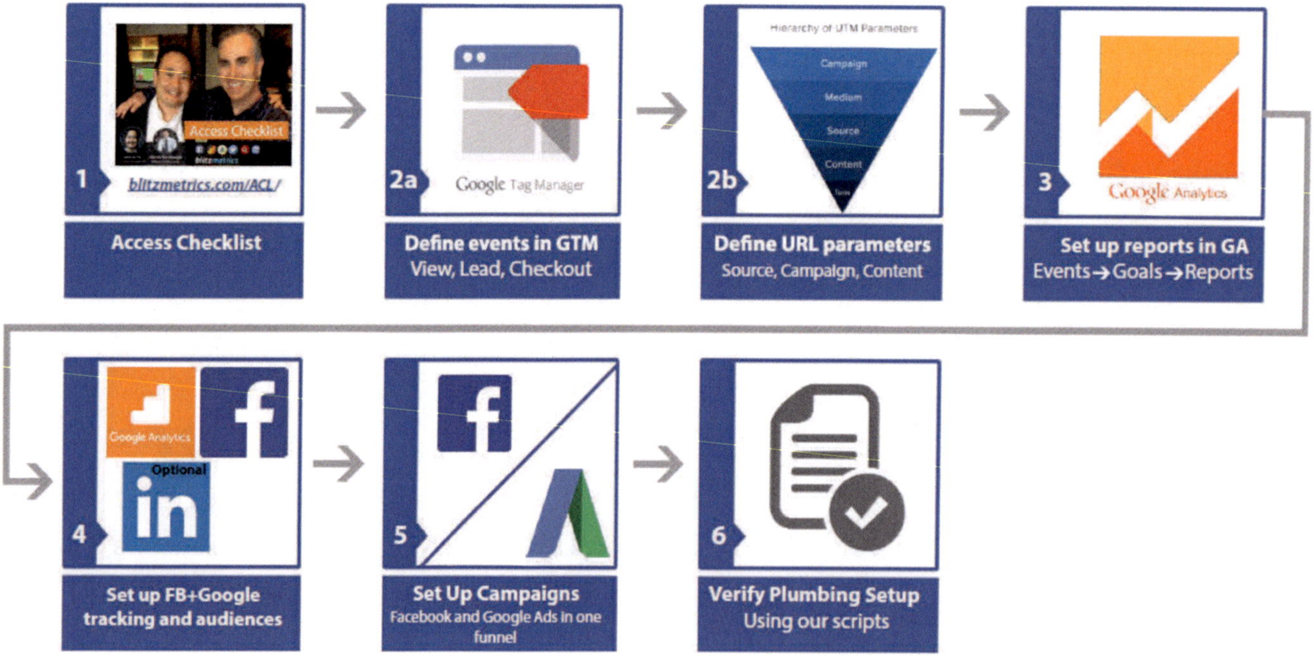

Setting Up Google Tag Manager with Google Analytics, Facebook Custom Audiences & Google Ads Remarketing

Creating Your Google Tag Manager Account

With Google Tag Manager, you can track all of your tags from Facebook, Google Ads, or anywhere else.

Go to **https://www.google.com/analytics/tag-manager/** to set up a Google Tag Manager account and follow the instructions provided on the site. Once you agree to the Google Tag Manager Terms of Service, you'll see the Tag Manager code that you will need to install on your site. **Paste the code into your website to complete the integration.**

HTML Sites: If your site uses HTML editor, then copy the code and open up your website's HTML page template. Find the code within your page template and paste in the Google Tag Manager code directly after and hit save. You will need to do this for all pages or page templates that you want to use Google Tag Manager with.

WordPress Sites: There's a useful **plugin for Google Tag Manager.** You can just copy your Tag ID which is shown next to your account name in the GTM dashboard or from the code itself after **"?id=" and then paste it into the plugin settings page.**

- Create your Facebook Ads account using Business Manager **business.facebook.com.**
- Create your Google Ads account and tie it to the Google MCC.
- Create your Google Tag Manager (GTM) account.

Create your Google Analytics account.

Setting Up Your First Facebook Custom Audience

1. Go to **business.facebook.com** and click on **Ads Manager.**
2. Click the Create Audience button and select to create a **Custom Audience.**
3. Select **Website Traffic** so that you can build an audience from your website visitors.
4. Configure the tag to target **anyone who visits your website** and set a period of 180 days. Make sure you also give it a name you'll remember such as **"All OutrageousDigitalcom Visitors".**
5. After creating your audience, you should be presented with your pixel. If you don't see it, click the gear in the lower left corner to bring up your pixel code.
6. Copy your **Facebook Audience pixel** and go to **Google Tag Manager** to create a new tag.
7. Select Custom HTML as the product and give the tag a recognizable name (**e.g. Facebook Audience)** by clicking the pencil icon next to the tag title.
8. Click Continue and paste in the Custom Audience code provided by Facebook.
9. Set the tag **to fire on all pages.**
10. Your tag is complete and you can either publish to make the changes go live or add another tag.

Setting Up Google Analytics with Google Tag Manager

1. Go to the ***Google Analytics website*** and either create an account or log in to your existing site to get your Tracking ID. Make sure you copy the Tracking ID rather than the full code from **Admin > Tracking Info.**
2. Go to your Google Tag Manager account and click to create a new tag with the Google Analytics Product type.
3. Select the **Universal Analytics** tag type and give your tag a name (e.g., Google Universal Analytics).
4. Click **Continue** and paste in your **Tracking ID** that you copied from the Analytics site. You will also want to make sure that you check the box to **Enable Display Advertising Features.**
5. Set the tag to to fire on **all pages** and click **Create Tag.**
6. Once your tag is complete, you can click to **Publish** the container or add another tag.

Setting Up Google Ads Remarketing With Google Tag Manager

1. Go to the **Shared Library** section of your Google Ads account and click to **View Audiences.**
2. Click to **View Audiences** and then hit the big red **Remarketing List** button.
3. Pick the type of users/categories you want to use.
4. Give your list a memorable name and ensure that the **Remarketing Tag** option is selected. You will also want to set a **Membership Duration of 180 days** before saving the list.
5. Once your tag has saved, you can click on the blue [Tag] link to bring up the code that you need.

6. Copy your Remarketing tag into a text editor so that you can copy specify detail from it.
7. Go to your Google Tag Manager account and click to create a New Tag
8. Select Google Ads as the Product type.
9. Select Google Ads Remarketing as the Tag Type.
10. Paste in your Conversion ID and Conversion Label from your Remarketing Tag.
11. Click to Create Tag and then you can either publish your container or add another tag.

Publishing Your Google Tag Manager Container

1. Make sure you have completed any tags you were building and click Publish
2. Google Tag will confirm the details that have changed since ou last published. If you are happy, click Publish Now.
3. Your tags will now be live on any pages where you have added the Google Tag Manager code.

Setting Up Facebook Conversion Tracking

1. Go to the Conversion Tracking menu item in your Facebook Ads dashboard and click to Create Pixel.
2. Select the type of conversion that you want to measure and give it an asily recognizable name.
3. Copy the Conversion Pixel code that Facebook presents you.
4. Go to your Google Tag Manager account and click to create a New ag of the Custom HTML product type.

5. Paste in the Facebook conversion code. If you look through the code, you will see two instances of "value" and "currency" in the code. If you want to track values in Facebook, you can replace these details with beyond the scope of this course).

6. Click to continue then set the tag to fire on **some pages**

7. Click **New** to create custom firing rule and complete the required details. You will need to select a variable to check which will probably be the Page URL and a matching rule. The simplest matching Operation to use is contains. If your conversion completion page was example.com/thankyou.html, then you could either enter "example. com/thankyou.html" or enter "thankyou.html". The benefit of only using a part of the URL such as "thankyou.html" is that it will match all pages which contain the value. For example, it would match both of the following links if you sold multiple products:

    - example.com/product/thankyou.html
    - example.com/service/thankyou.html

8. Click **Save** to confirm the new rule and then **Create Tag** to add the conversion tag to your container.

9. Your **New** tag is now complete and you can either publish to activate the tag or add another tag.

Setting Up Google Ads Conversion Tracking

1. Go to the Tools > Conversions section of your Google Ads account and click the + Conversion button.

2. Click to add a new Conversion item and then select Website as the source of conversion.

3. You only need to set a memorable name for your conversion tag as you can leave the rest of the details to Google Tag Manager so click Save and Continue.

4. Copy your conversion tag to a text editor so you can copy the Conversion ID and Conversion Label later.

5. Go to your Google Tag Manager account and click to create a New Tag of the Google Ads product type.

6. Give the tag a memorable name and select Google Ads Conversion Tracking as the tag type.

7. Find your Conversion ID and Conversion Label in the code you copied from Google Ads earlier and then paste them into the tag configuration. You can also set a specific conversion value and currency here, or use dynamic variables (advanced method beyond the scope of this course).

8. Click to continue and set the tag to fire on some pages.

9. Click new to create a custom firing rule and complete the required details. You will need to select a variable to check which will probably be the Page URL and a matching rule. The simplest matching Operation to use is contains. If your conversion completion page was example.com/thankyou.html then you could either enter "example.com/thankyou.html" or enter "thankyou.html".

10. Click Save to set your new firing rule and then Create Tag to complete the setup.

11. Your new tag is now complete and you can publish your changes or add another tag.

Google Tag Manager Tips Version Control

Once you publish a container, the existing version goes live onto your website and the version shown in your Google Tag Manager editor becomes a draft. You can then edit this draft freely - without changing the code that is being served on your live site - until you hit Publish again.

Managing Users

You can add more users (e.g. agencies or web developers) to your Google Tag Manager account in just a few clicks.

1. Go to the Admin section of your Google Tag Manager account and click User Management.
2. Click New to add new user and then enter the email address of he person you would like to invite.
3. Select the features that you want to grant access to under Container Permissions and then Click Confirm.

Verifying Your Tag on the Published Site

To verify that Google Tag Manager is installed on your site, you can install the Tag Assistant extension to the Google Chrome browser and visit your website. If everything has worked, you will see a green icon in your toolbar which you can click on to view the details about your tag.

Setting up Twitter Remarketing Audiences

1. Go to your Twitter Ads Dashboard at https://ads.twitter.com/.
2. Input payment information.
3. Select Tools > Conversion Tracking.
4. Click to create your first website tag
5. Enter a descriptive name for your first audience set a conversion type of Site visit while ensuring that the tailored audience box is checked.
6. Click to Show Conversion Settings and set appropriate tribution windows. For a simple catch-all remarketing tag, we recommend 90 days post engagement attribution and turning Off or setting a low 1 day post-view attribution..
7. Click Save tag and generate code snippet and you will be presented with your remarketing tag, which you can now add to Google Tag Manager.

Adding your Your Twitter Remarketing to Google Tag To Google Tag Manager

1. f you haven't already got your Twitter code on hand, you can access existing tags under the Tools > Conversion tracking page by hitting the Edit option on your audience.
2. Go to your Google Tag Manager account and click to Add a new tag.
3. Enter a descriptive name for your Twitter Remarketing Tag and select Custom HTML Tag as the type.
4. Paste in the code that you got from your Twitter Ads account and hit Continue.
5. Click the All Pages option to ensure your primary remarketing site catches all visitors to your site, and then click Create Tag.
6. Click Publish and then confirm with the Publish Now button in the box that pops up.

Adding Your Instagram Account to Your Facebook Advertising Account

1. Go to Facebook Business Manager at business.facebook.com.
2. Hover over, or click on, Business Settings and then select Instagram Account.
3. You will usually be presented with a list of Instagram Accounts but as you haven't linked one yet, click to Claim New Instagram Account.
4. Enter your Instagram Username and Password to the pop up box and click Next.
5. You will be presented with your linked Instagram Accounts, showing the one you just linked. Click to Assign Ad Accounts, select your Facebook Ads account and click Save Changes.
6. (Optional) You can also click to Assign Partner to allow an agency to manage your account. You can then simply enter the Business ID they provide.

Goals

- ✓ Define your mission, start with (WHY) and identify desired outcome and customer segments.
- ✓ Identify your primary goal(s) in the next 90 days.
- ✓ Determine your target Cost per Acquisition (CPA) or Return On Ad Spend (ROAS).
- ✓ Determine your ads budget relative to campaign goals (optimizing for clicks, page likes, form submissions, etc.).
- ✓ Choose one key metric for each funnel stage: Awareness, Consideration, and Conversion (ACC).
- ✓ Develop your brand via the 6 phases of the Personal Branding.

☐ Fundamental ☐ **Advanced**

Laying down the plumbing for an awesome article machine

Figure out what you stand for (your "WHY") want to brand yourself as. If you're passionate about music, you might want to become the best Facebook marketer in the music industry. Or maybe you're passionate about sports. Establish yourself as the top social analytics expert in the sports industry.

Once you figure out what you stand for everything else falls into place.

Writ copy that fits into your content pyramid.

The most important thing to remember when writing your own content is to make sure it all connects back to your "WHY". Even if your "WHY" is something general like content marketing, you can break that down into storytelling, promotion, writing techniques, etc. If you continually break down your **"WHY",** you'll never run out of topics over which to create content. Through the process of Learn, Do, Teach (one of our **"9 Triangles")** as you're creation opportunity. This could be as simple as sharing a little tidbit you found interesting in a book you were reading, or a longer, more thought out strategy analysis.

Checklist Creation

If you take what you've learned and applied within your subject of expertise and condense it down to simple steps, you can assemble them into checklists. This is your big selling point. Except you can't sell it. . This is where people get tripped up the most, and for good reason. You probably feel like you'd just be giving out your secrets for free and letting other people piggy-back off of your hard work with no monetary return on your end. By offering these check list it shows you know what your talking about that you are a expert in your field. How else could you condense down all this information that would normally take someone thousands of hours to go through into a simple, one page checklist?

Internal support from Facebook

If you build your relationship with Facebook, your representative can run ad account audits for you, outlining key metrics and breakdowns, with actionable items for you to apply. Use this sparingly and for accounts that spend north of $100k a month.

HC Report	Use Case	Objectives	Primary Metrics	Products Included	Breakdowns
Direct Response					
DR Solutions	Snapshot view of DR activity	• Website Clicks • Conversions • Product Catalog Sales	• Cost per Conversion	• Lead Ads • Carousel Ads • Dynamic Ads • FB Pixel	• Bid Type • Day Parting • Demographics • Placement • Targeting
App Ads	Performance of a particular app ad	• MAE • MAI	• CTR, CPI, CTI • Competitive Benchmark	• Mobile App Install Ads	• Ad Type • Audience • Bid Type • Device
Mobile App Installs	Performance of mobile app installs	• MAI	• CTR • CPI • CTI	N/A	• Audience • Bid Type • Demographics • OS • Placement
Mobile App Engagement	Performance of mobile app engagement	• MAE	• Mobile App Action Rate • Cost per Action	N/A	• Audience • Bid Type • Demographics • OS • Placement
Brand					
Video Insights	Snapshot view of video performance	• Video	• Video Views • Video View Rate	• Video Ads	• Demographics • Retention
Brand Solutions	Evaluation of brand delivery with aim of driving more brand option	• BAO • Reach & Frequency • Video	• Reach • Impressions • Top Campaigns & Ads by EARL	• BAO • Reach & Frequency • Local Awareness	• Audience • Bid Type • Demographics • Placement
Business Review					

What was the last thing the customer did before they clicked?

There can be multiple clicks by a single person.

Triggers sequences of events according to actions by customers (think IF/THEN).

Sequences (spiral) - More and more sequences create a spiral efforts.

How to get from point A to B. Try another path to find out how else to get to B. Test to find out the better conversion rate, etc.

Media mix modeling - Different channels impact or different cities. What type of media creates more impact?

1. **Last Click,** The most common, but flawed model, that gives 100% credit to the last thing the customer did before conversion. Penalizes the "assist"or word of mouth power in other parts of the funnel. By default, you're doing this.
2. An **Overlap** is when people are part of two audiences. They could like pizza AND chicken wings. An Affinity Grid shows the set of interests and behaviors that describe our target customer. Actions can be overlaps, too, if they have been on the site in the last 7 days and are a fan. This lets us calculate propensities to convert.
3. **Triggers,** unlike overlaps, require order of operations. IF/THEN statements (Event A happens and then Event B). An overlap helps us initially determine when two targets are related, but a trigger tells us which target comes first. Branded search is triggered by something upstream, for example.
4. **Sequences** are triggers stitched together. The most complex of funnels can be disassembled into atomic building blocks of triggers. The most powerful sequences go cross-channel from social to email, from email to site, from app to in-store visit, or other combinations.
5. **Lift Testing** measures the incremental impact of a marketing action. This variants. Lift testing requires a holdout audience (the control of a test and control experiment) so we can measure true incremental impact. We can randomize by many possible variables.
6. **Media Mix Modeling** is the most sophisticated and final stage of analytics. It allows us to see how all channels work together to drive conversions in the customer journey. All prior phases are pre-requisites.

GOAL SETTING CHECKLIST

In one sentence, describe your business's mission. In the sentence, identify the desired outcome and your customer segments.

e.g. Provide young adults with the opportunity to turn their career and life aspirations into reality.
f you have define leads or sales as one of your goals what is your target cost per action or return

If you have defined leads or sales as one of your goals, what is your target cost per action or return on ad spend? (Use historical data and a theoretical calculation of your break even cost)

Cost per Lead: $ ____________

Cost per Sale: $ ____________

Target Return on Ad Spend (ROAS): ____________ %

How much can you comfortably spend monthly to test and learn?

☐ $100 ☐ $500 ☐ $2,500 ☐ $10,000+

What is the value of your fans?

There are three ways to calculate this: conversion value, earned media value, and comparable's. For the purposes of this exercise, we've chosen comparable's to score your Facebook page because we are using publicly available data to see whether you have engagement apps and to what degree you are interacting with your fan base.

Method #1: Conversion Value

The most accurate valuation is to use your own conversion data to determine what percentage of Facebook fans perform an action with an assigned dollar value, such as a shopping cart check-out, filling out a lean form, joining your email list, or other activity.

If you are a direct marketer, consider what a conversion is worth and what percentage of fans convert. It may be that your Facebook fans will learn about you first on Facebook and then go to Google prior to converting. If that's the case, and you're using a last click attribution model, the Google search gets 100% of the credit, leaving no credit for the "assist" from social media. **Do your analytics show you how your various marketing efforts play together to drive a conversion?** If not, you may be over-awarding credit in the same way that an analytically minded retail client could accidentally conclude that, because cash registers are correlated with sales, they should divert marketing dollars into installing more cash registers in the store.

The beauty of the e-commerce and lead gen model is that it's simple to measure the ROI of your Facebook presence. You need only conversion tracking that accounts for the impact of ALL marketing efforts, not just the last things a customer did prior to conversion.

Method #2: Earned Media

If you are a **Consumer Packaged Goods** brand or one that interacts with customers via retail channels, then earned media value is most accurate to determine value.

You can measure how many impressions your Facebook page is generating. For example, if you are generating **100,000 impressions per day** via your **Timeline posts,** and your cost of media is a **$5 CPM** in the display, then your Facebook page is worth $500 per day. That's what you would have paid to generate the same level of traffic.

Arguably, the **Facebook traffic is worth more** because it is trusted, has a higher engagement rate, and allows for sharing. What if you only post once per week and your **Post Quality Score** is such that few people notice your comments?

You might have **1,000,000 fans** but be generating only a **few thousand impressions** a day.

In finance, if you want to buy a company, there are three primary valuation models you would use: **discounted cash flow, comparables,** and **net asset value.** The **first (DCP)** is a lifetime of projected earnings discounted back via an implied interest rate.

The second is to say what similar companies are worth and apply their same p/2 ratio or composite set of metrics. For example, if Google is worth $137 billion and has 60 percent share, then Yahoo!, at 5 percent, should be worth 1/12 that market cap. The third is liquidation value- if you sell off all that company's assets and pay off all debts, what are you are left with?

Each technique is appropriate in the right situation. For example, you wouldn't use NAV with software companies, since they don't have assets (and goodwill via acquisition doesn't count).

The same three valuation techniques apply to Facebook fan value. Let's step through them.

1. Discounted Cash Flow (DCF) is the most revered technique since it's the closet to real profit and not subject to stock market manias. Thus, Outrageous Digital can pixel the conversion page of a website to capture not only the conversion but whether they are a Facebook fan or not. If so, assign credit. Then calculate what percentage of conversions are Facebook fans and what the average conversion value is for Facebook fans vs non-Facebook fans.

Now, you have a real Facebook fan value by taking checkout value of the Facebook fan times what percentage of Facebook fans buy (survey randomly on all site users who are fans or not compared to purchasers who are fans or not). Indisputable value can then be plugged into our grading tool- to override the default fan value. Some people say that pixeling is easy, and I know that we built a "popup-like" survey tool that can gather this via a light-box JavaScript. Of course, this technique is highly flawed because correlation is not causation, and we don't know lifetime value. Few marketers realize this; they assume that Facebook users are more valuable implicitly because they truly are higher income, early adopters, iPad users, etc. Let's say we released a study where we found that those people who wear ties are more likely to speak at a conference.

2. Industry Multiples is the trickiest of valuation techniques because it's subject to speculative bubbles. Just because everyone thinks a fan is worth $10, then you drive off the cliff, too. ClickZ published a "study" showing fans to be worth $3.60 as if all fans are homogeneous. Adidas said fans are worth $90 based on the faulty logic that these people who are Facebook fans wouldn't have bought otherwise; they didn't measure incremental value but took full credit.

There is no lifetime value of a Facebook fan because nobody has perfect vision into what changes Facebook will make, what the shift in marketing channels will be over time, and so forth. So it's tricky to compare against direct mail, TV, or other more stable media. Perhaps the closest is email, which is semi-interactive and an initial acknowledgment of permission-based marketing. I've used 2x as a rule of thumb- to say that a Facebook fan is worth 2x an email subscriber. Why 2x? Because emails go stale faster - people check their Facebook often multiple times per day and aren't about to discard their identity for a new one each month. They don't have multiple Facebook accounts. And there is no social power in email, since replies are not broadcast into the feed. The recommendation power of a Facebook fan, properly leveraged, is tremendous.

3. Asset Value

1) If you were to sell your list to others, what would that fetch? If you were to sell ad space on Facebook (not allowed, of course), how much could you generate? If you were to buy the same amount of exposure via a display campaign, what would that cost you? That's called earned media and is often based on relative CPMs or what you are paying

Google for clicks/visits.

2) What would it cost to buy the traffic? Here, we are assuming that markets are efficient and that your advertising prowess is middle of the road. If you spend $50k on Facebook ads, it must be worth that in an efficient market. Not a good valuation model, but it's actually the most common model to value advertising. Don't believe me? Almost all agencies charge a fee based on media spend. So if you have an agency, then you are buying based on assuming the price is the determinant of value. Even worse is buying on a rate-card, which is the equivalent of buying at "list price", negotiating a 15 percent discount, then claiming it was a good deal.

When you talk to advertisers, know which model they are using. Agencies understand the percent of spent markup. Performance players (lead gen and e-commerce) look at the firsy model (ROI and its cosin NPV). VC funded companies use real currency, so use the second model.

Measuring Engagement

If you're an administrator of your brand's Facebook page, you can access Insights. Here are three key metrics to determine your engagement:

1. Feedback Rate

What percentage of your fans are interacting with you? Is it under 1%? If so, then your fans are saying your messages are not interesting. Facebook, noticing this, will also not share your messages with friends — a double whammy. Improve your score here by asking questions as opposed to making statements, keeping messages 10 words long or less (versus a paragraph), including multi-media versus just text, and having content that triggers emotion. Above all, resist the temptation to sell too hard. The Timeline is where most engagement occurs, although you can boost your engagement rate by running ads to your postings, plus including engagement apps.

2. Interactions Per Week

While Post Quality Score measures the share of your audience that is engaged, Interactions per Week looks at how many interactions you have. Under-post, and you have no community. Over-post, and you wear users out. Your ideal post frequency is dependent upon what your users expect. If you're a news site, then 10 times a day can be okay. If you're a B2B software company, afew times a week is probably fine. Do not cheat by auto-posting your RSS and Twitter entries—fans will frown upon non-social, automated content. Don't promote your specials or discounts more than once in every four postings, and definitely no more than once per day. If you are getting less than 10% of your fan base as interactions each week, you're not unlocking the value of your community. Thus, if you have 100,000 fans, you should have at least 10,000 interactions each week.

3. Troubleshooting Low Engagement

We encountered low engagement with a hair care products company who grew their fan base by running contests and then using the Timeline to push their latest products. Here is how we explained the problem to them:

"Contest traffic is the lowest quality - people who come in via incentives are not there because they truly care about shiny, frizz-free hair. The worst quality traffic is pure game incentive traffic (where people do something just to earn coins). Our traffic is somewhere in the middle. It's reasonable to expect a 10-20% loss once the promotion ends. That's just the price you pay for growing quickly via ad-driven traffic that runs to contests."

"Clearly, the highest quality traffic is viral (organic) growth from friend driven engagement. I'm not talking about game notifications or incentives for inviting friends, which is the spam behind the FarmVilles of the world. I'm talking about testimonials and real participation. We don't have such apps but can construct them with a moderate amount of effort. This includes a 'spin the wheel' coupon app (which is better

than just loudly broadcasting some special or discount), a 'featured fan' (nomination) app, that shows folks in the profit image, and some game-like element(we'd have to see how this is possible - maybe using video or sharing bumper stickers/slogans)."

"Thus, certainly based on what we have so far - all ads and no further engagement except Timeline postings, we're going to lose many of our new fans - many naturally and others because we don't have ways to participate."

Business Objectives by Benchmark

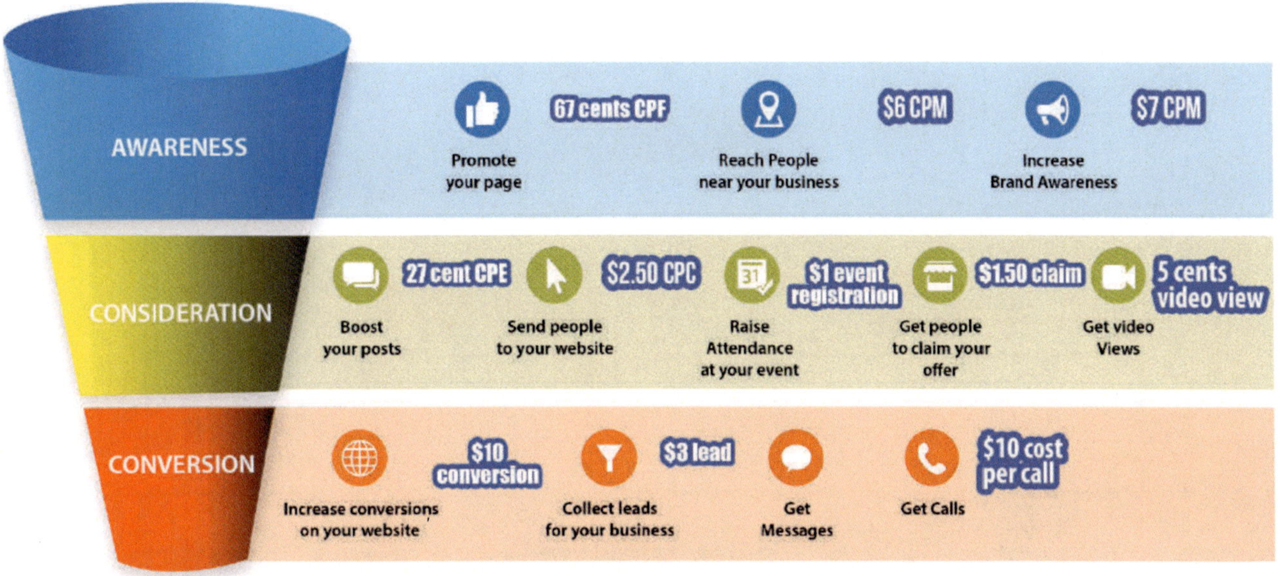

Metrics Decomposition

Find out exactly where the problem is with your funnel by traversing the Metrics Decomposition Pyramid. You can trace down the particular metric that's hurting your conversions, then take the right action.

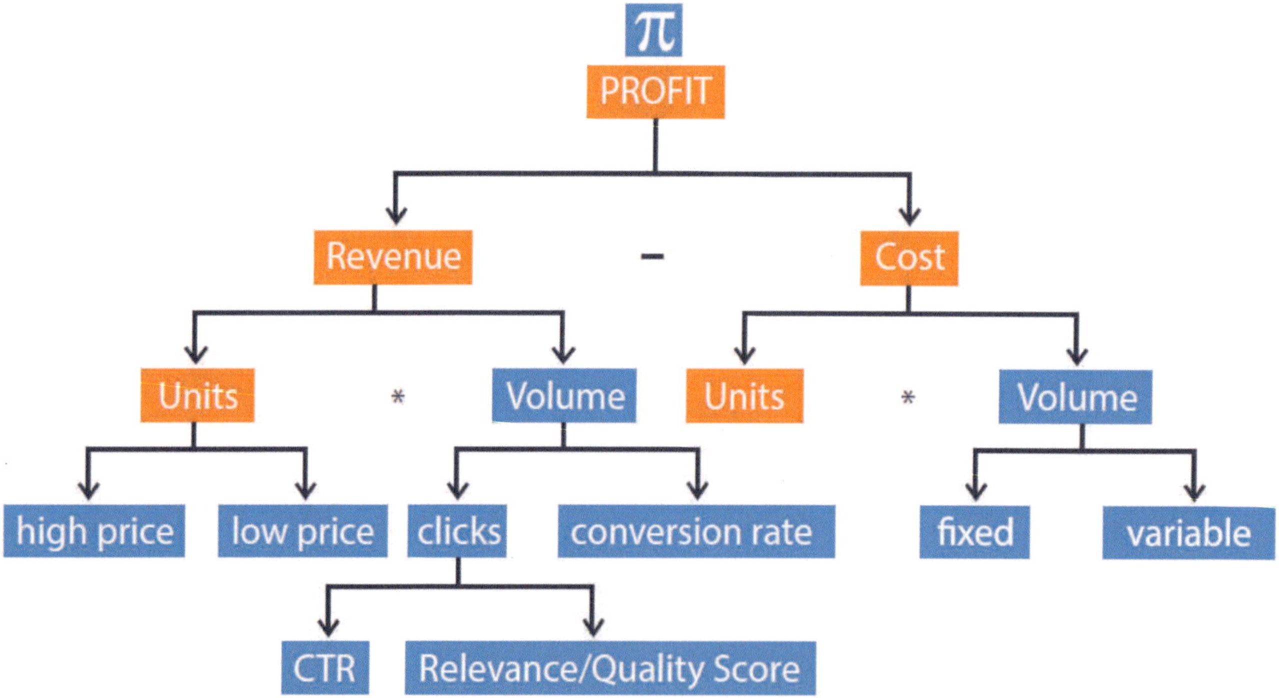

We can set up automated rules on almost every measurable metric within Facebook. Find your business's weak point(s) and make sure that you have an automated rule set up to kill any that are draining our ad spend. For example: *Our Standards of Excellence CPM says we should aim to pay $10 or less for 1,000 Impressions. To ensure we stay within this, let's have a rule set up which kills any ads which are driving CPM's higher than $15***.**

We can set up automated rules on almost every measurable metric within Facebook.

Find your business's weak point(s) and make sure that you have an automated rule set up to kill any that are draining our ad spend.

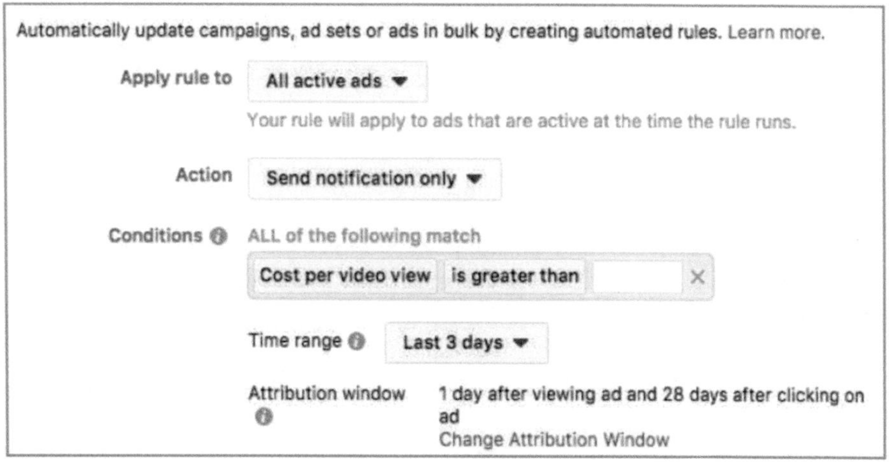

For example: Our Standards of Excellence CPM says we should aim to pay $10 or less for 1,000 Impressions. To ensure we stay within this, let's have a rule set up which kills any ads which are driving CPM's higher than $15.

You can choose varying different degrees of automation.

Increasing Spend, Decreasing Spend, Turning Ads Off and just Spending a Notification.

If you are slightly unsure of the automation, have Facebook send you an email notification when your conditions are met. This way, you can jump into the ad

account and adjust your bid manually.

Once you feel comfortable with the automated rules, you can have Facebook adjust your bidding/ads automatically.

Here are a few rules you can get set up today:

Send Notifications of High Ad Frequency.

Apply Rule to All Active Ads >> Send Notification Only >> When Frequency is greater than 3.

Send Notification of Low CPCs, so we can increase budget. Apply Rule to All Active Ads >> Send Notification Only >> When CPC is below $0.75.

How to Set Up your Automated Rules

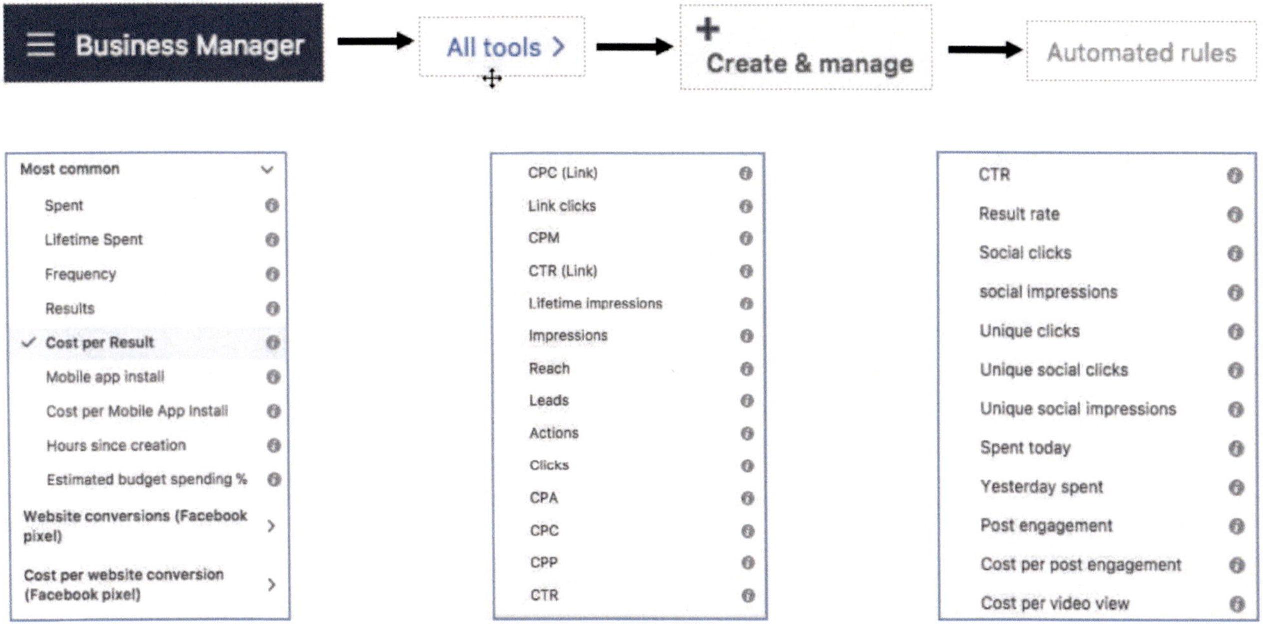

Content

- ✓ **Assemble a list of third-party endorsements, especially positive mentions from high-authority sites.**
- ✓ **Create a 3-minute "WHY" video Set up Content Library.**
- ✓ **Map out one-minute videos.**
- ✓ **Create promotional content to drive conversions.**
- ✓ **Create personas for the top 3 customer segments you serve.**
- ✓ Gather content by stage of the ACC funnel, addressing a key objection for each persona.

☐ Fundamental ☐ **Advanced**

- [ ] What is your story? About page, write/copy and paste the link in the blank below).
  Why did you start this business? e.g. family legacy, survival, personal experience, friendship, etc.

- [ ] What is your business the BEST in? What niche do you aim to dominate? *e.g. Nike may be the biggest sports apparel brand but Under Armour*

- [ ] Why do people love you? (If information about your Unique Selling Proposition is published online, write/*copy and paste the link in the blank below).*
  What are the reasons why your customers buy from you rather than your competitor? What is your unique selling proposition?

- [ ] What credible proof do you have to tell your story? Is the proof online? Can you share them
  What are some examples of your unique selling proposition? e.g. customer testimonial for a cross-fit trainer?

- [ ] Who produces content for your business?

- [ ] How many videos do they produce in a week?

- [ ] Do you collect positive and negative mentions?

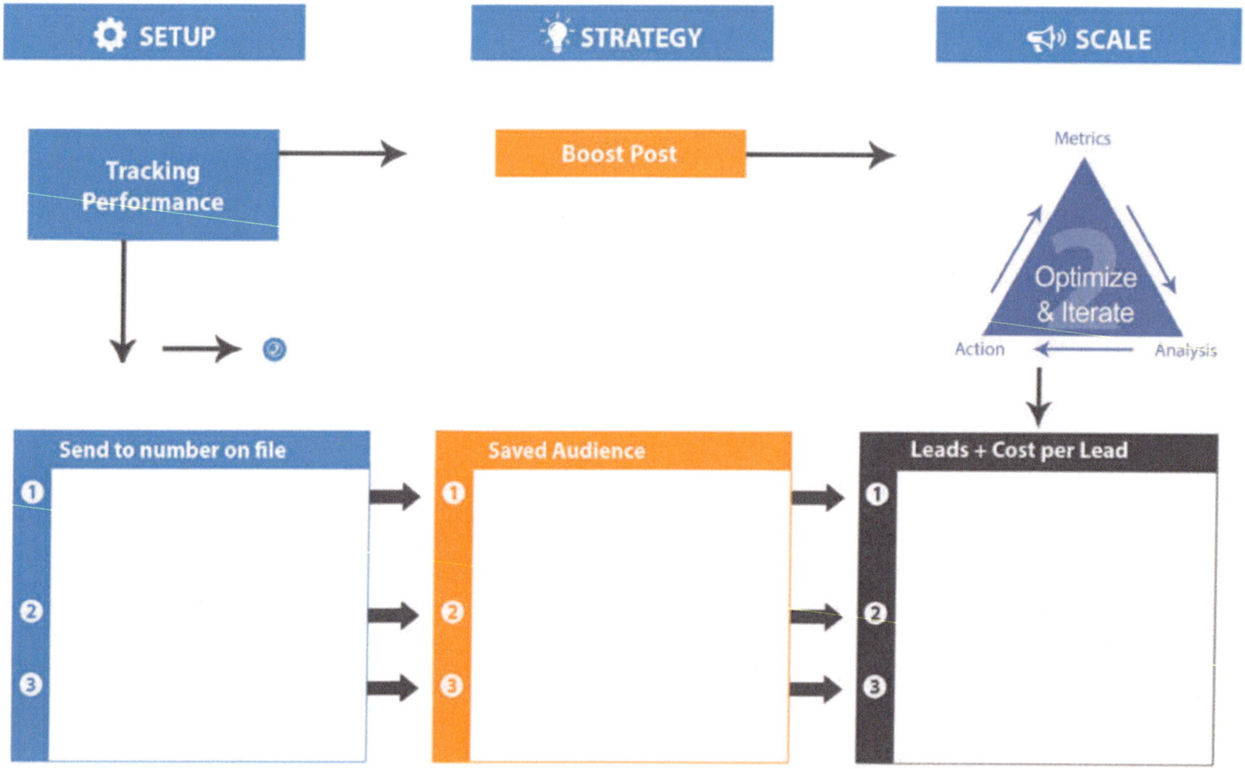
SETUP
STRATEGY
SCALE
Tracking Performance
Boost Post
Metrics
Optimize & Iterate
Action
Analysis
Send to number on file
Saved Audience
Leads + Cost per Lead
1
2
3
1
2
3
1
2
3

Facebook Lead Gen Machine

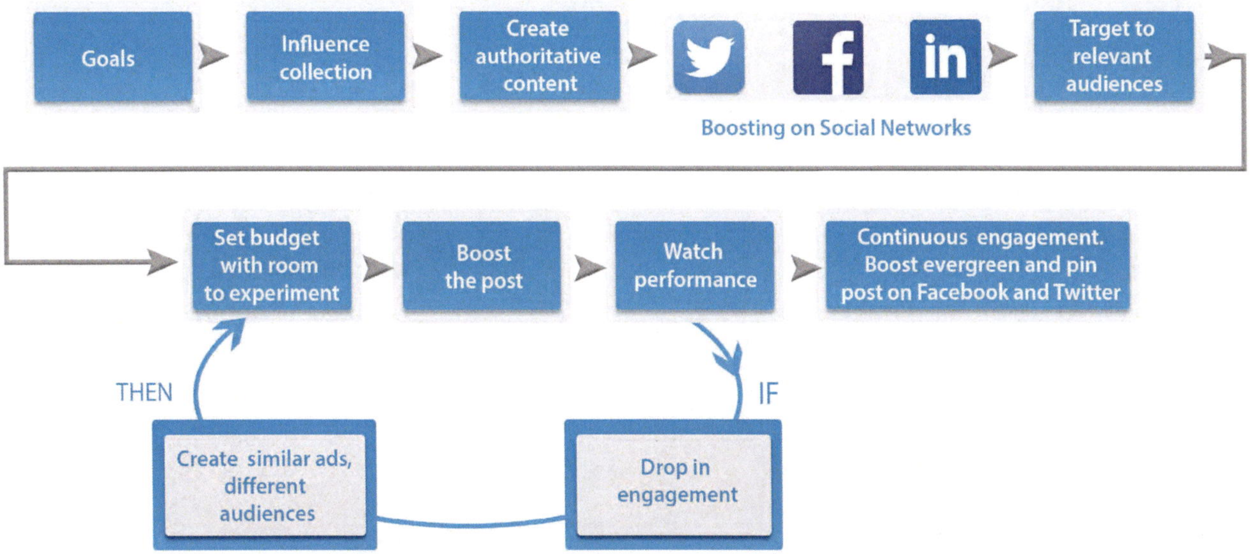

Dollar a day Layering Technique

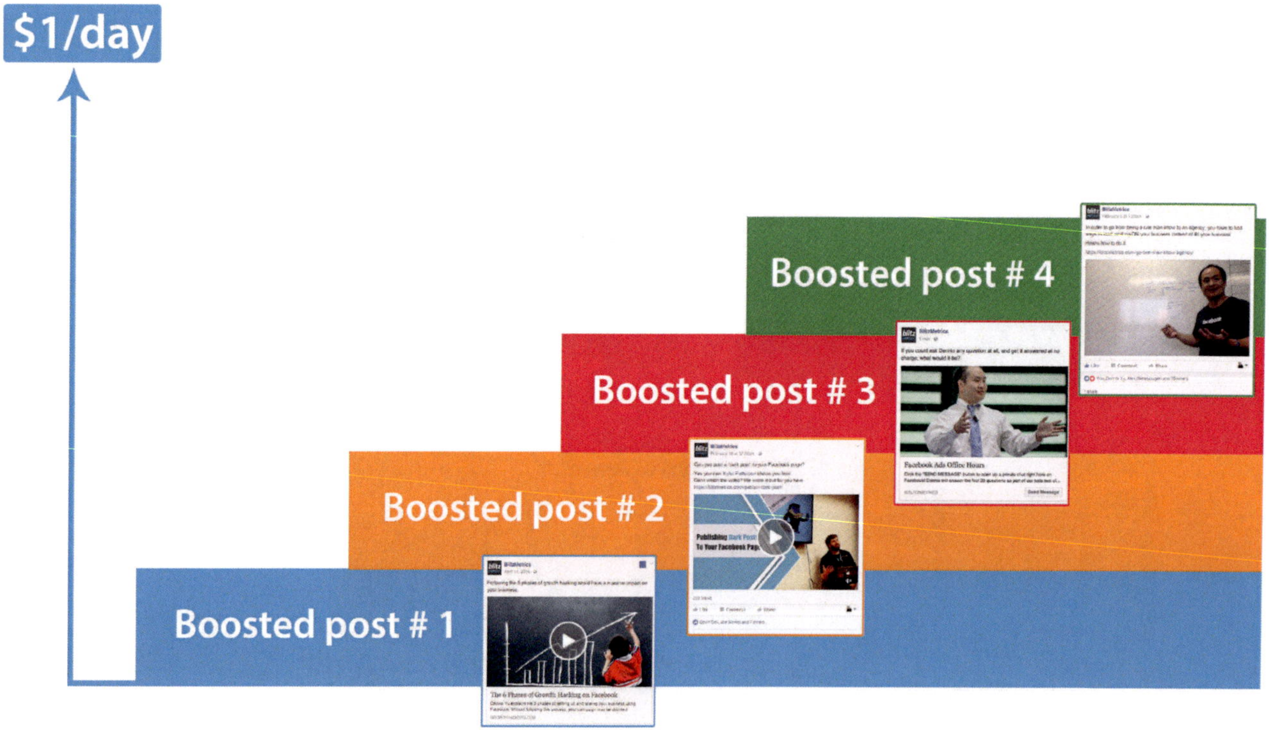

Dollar a day Layering Technique

Personal Branding

WHY "dollar a day"?

- Makes any goal easier to achieve
- Build authority by influencing the influencers

Why Story 3 Components

- Valuable experience
- I believe that...
- And because of this, I do...

WHAT is "dollar a day"?

- Generates inbound marketing
- 3 linked phases that multiply your power

Plumbing

- Website & social accounts
- Professional & consistent branding
- Tracking & ad pixels

Content

- Third-party authority
- Content library
- Collection mechanism

Amplification

- Creating influencer targets
- Dollar a a day ads per post
  Responding to feedback

HOW to use "dollar a day"?

3 Components of authority

- Properties
- People
- Content

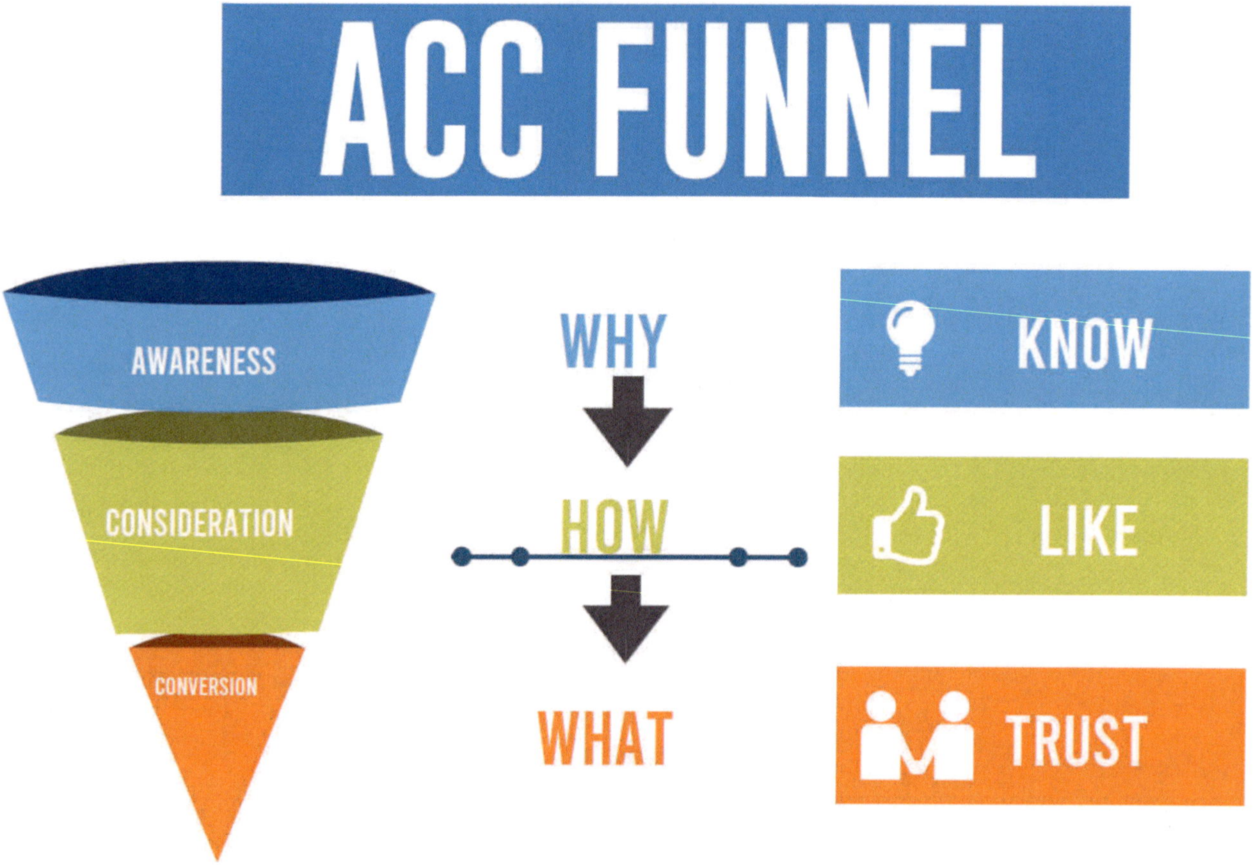
ACC FUNNEL
AWARENESS
CONSIDERATION
CONVERSION
WHY
HOW
WHAT
KNOW
LIKE
TRUST

3X3 VIDEO GRID

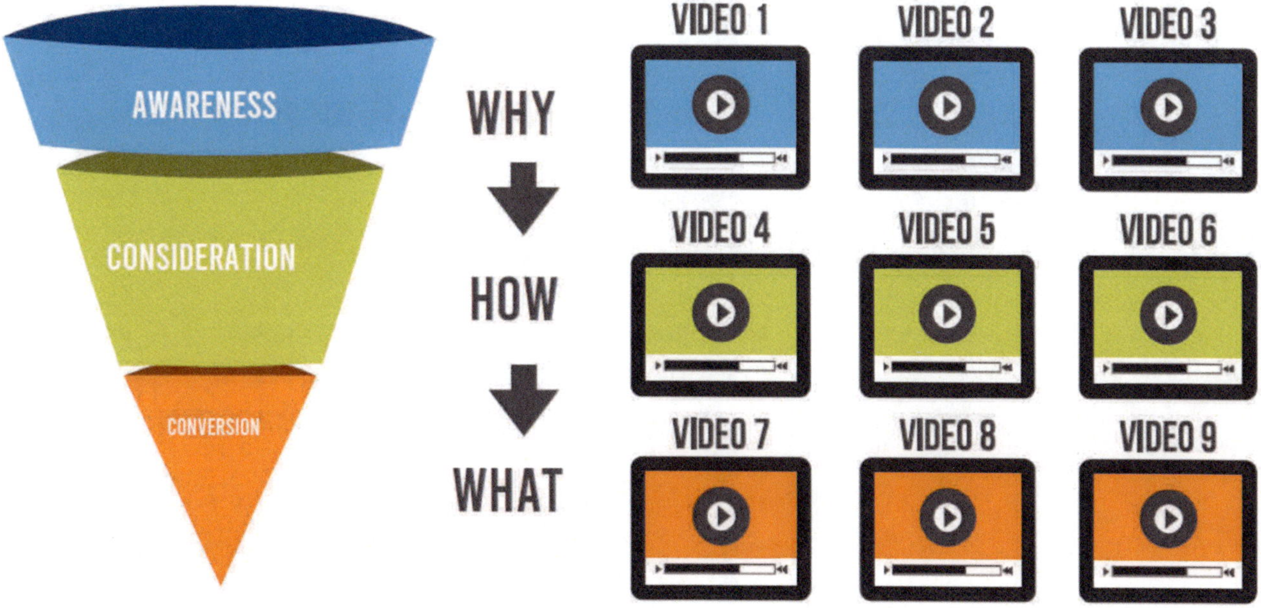

FUNDAMENTAL 3X3

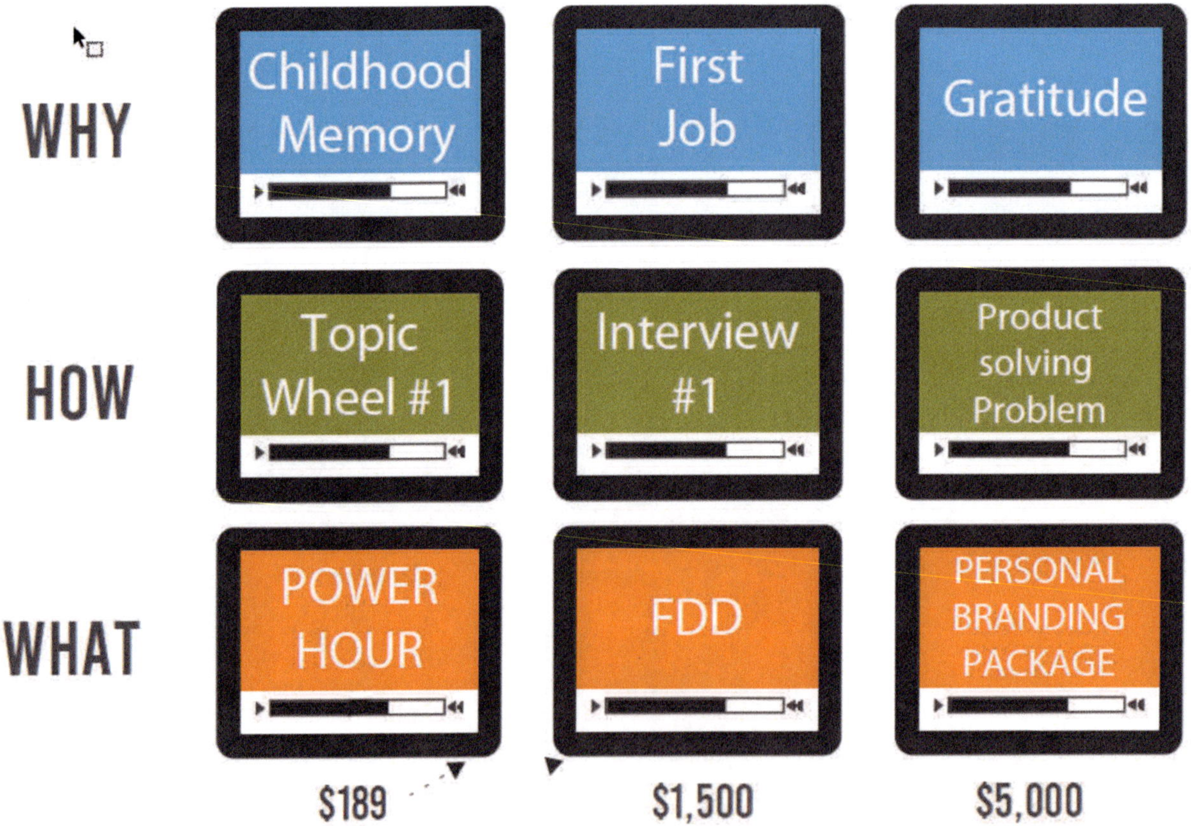

When is video HOT? 4 signals

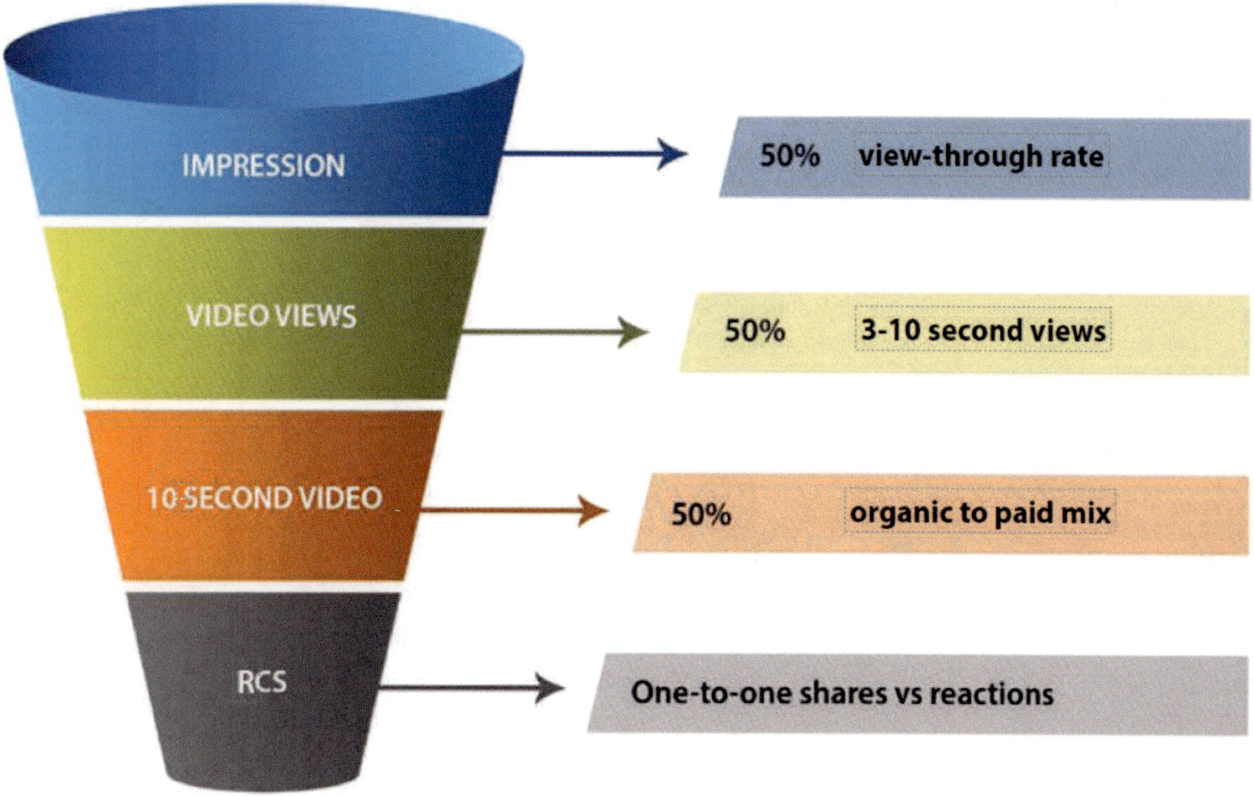

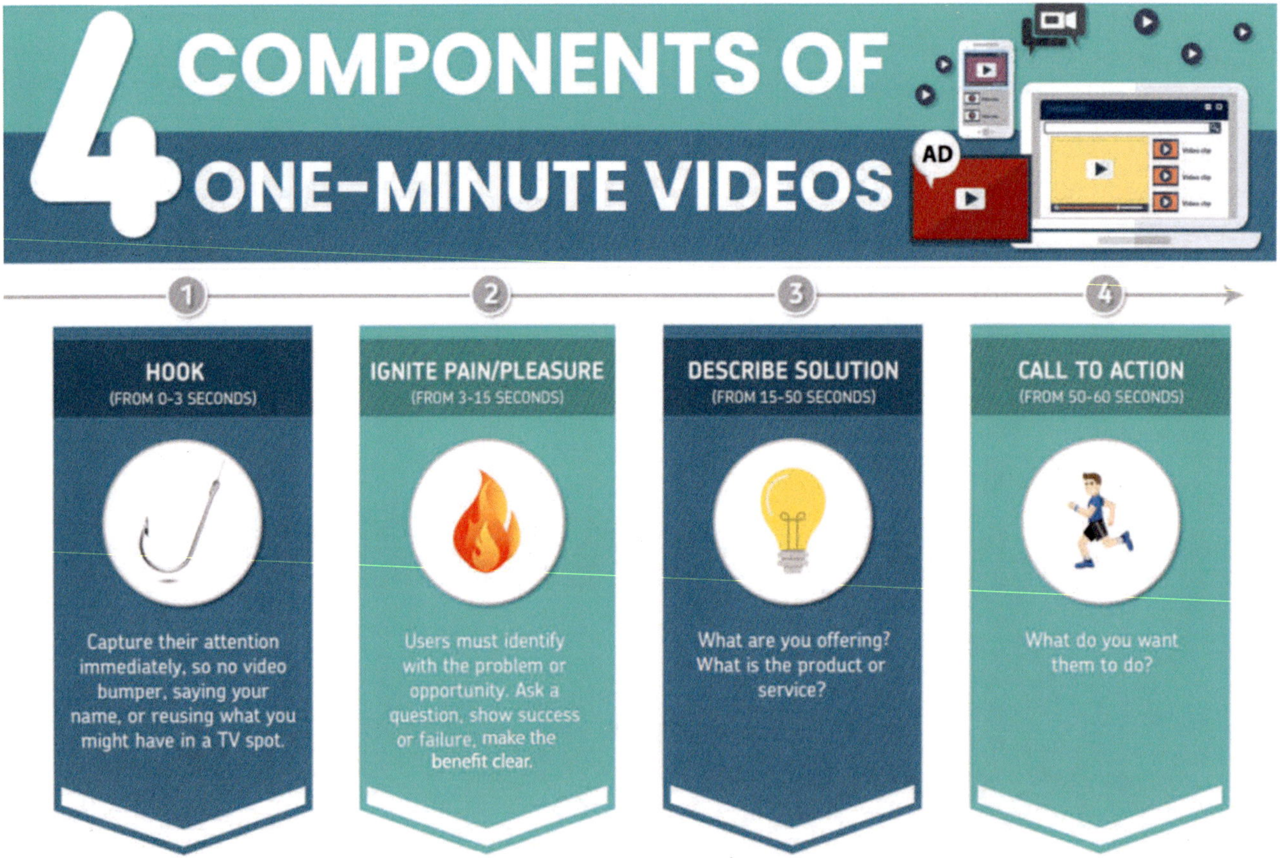
4 COMPONENTS OF ONE-MINUTE VIDEOS
AD
1
2
3
4
HOOK
(FROM 0-3 SECONDS)
Capture their attention immediately, so no video bumper, saying your name, or reusing what you might have in a TV spot.
IGNITE PAIN/PLEASURE
(FROM 3-15 SECONDS)
Users must identify with the problem or opportunity. Ask a question, show success or failure, make the benefit clear.
DESCRIBE SOLUTION
(FROM 15-50 SECONDS)
What are you offering? What is the product or service?
CALL TO ACTION
(FROM 50-60 SECONDS)
What do you want them to do?

3 COMPONENTS TO A WHY VIDEO

1. WHEN I WAS___.

Waste no time; get right to the story. This is where the "WHY" comes in.

2. I BELIEVE THAT ___.

Go from the emotion of the story you just told to the overarching lesson of what you stand for.

3. I STARTED ___.

Give a brief explanation of what you do. Make sure to be specific and concise.

INSTRUCTOR USE

- [ ] As each person comes up, let's clap for them-- it's own slide of "cheering!"
- [ ] As they hit each of the 3 components, we advance to the each slide, so they know what trigger phrase to use.
- [ ] Then we go to a slide for "cheering!" so people know to clap as they leave the stage.
- [ ] And finally, "feedback" slide, where the instructor(s) provide 2 minutes of feedback before calling up the next participant.
- [ ] When we call up the next participant, we quickly left arrow 5 times to get back to the first "cheer!" slide.

One million video views for $65

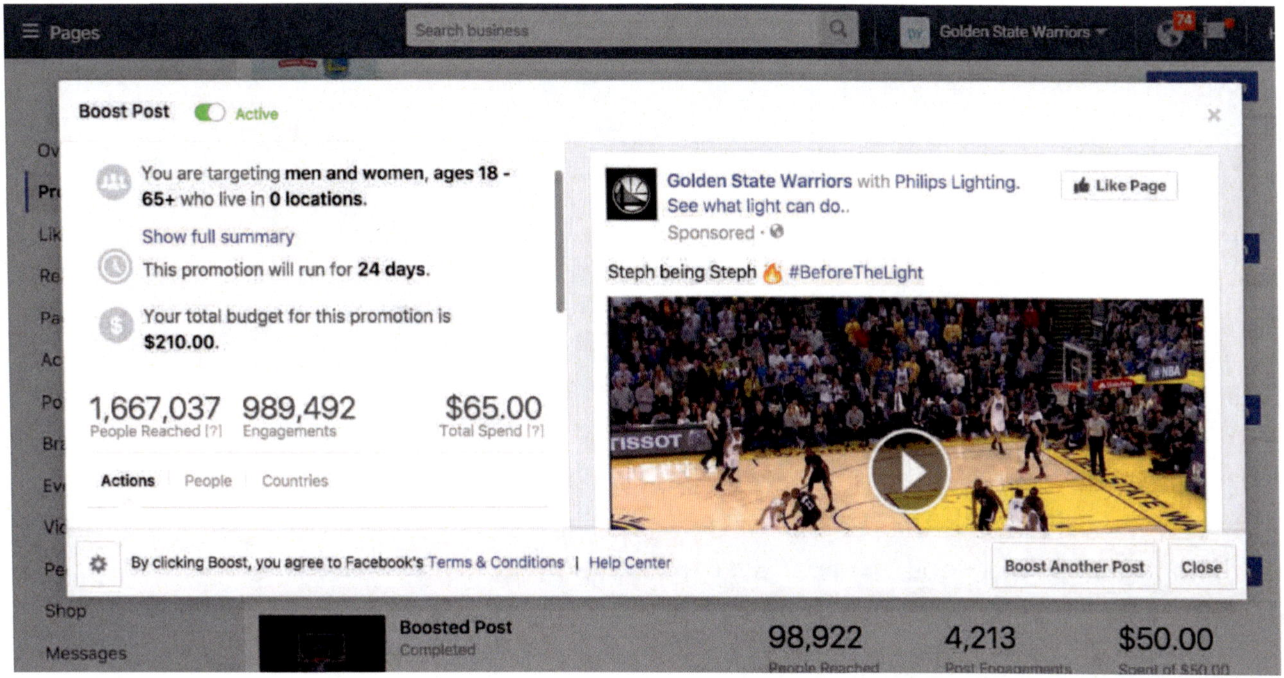

Promote video when your audience is most engaged, such as this post during the NBA playoffs. You are going to get the best results when your fans are the most engaged. This video boost got 1,600,000 views for $65. That's our tenths of a penny for each view. When you are posting your content and promoting it, try to capitalize on the best time and occasion.

Section 1: Set up your plumbing

Define what you stand for (your "WHY")

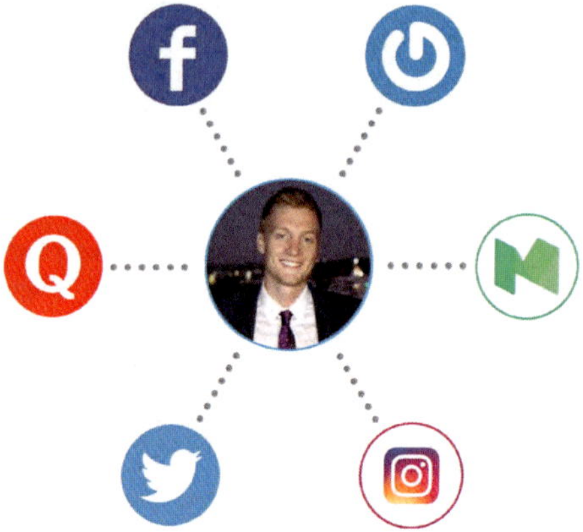

Before you start the process of personal branding, you need to define what you want to brand yourself as.

If you're passionate about music, you might want to become the best Facebook marketer in the music industry.

Maybe you're passionate about sports. Establish yourself as the top social analytics expert in the sports industry.

Once you figure out what you stand for, everything else can start to fall into place.

It's the through-line that drives the content you produce. This is critical to inbounding the clients you want to work with because people hire people.

While it's important to look good on paper, even more important to potential clients and employers is whether or not they'll like working with you.

People spend the majority of their days at work, so of course, they want to hire someone personable whose goals, values, and personality aligns with theirs. The same goes for public figures.

If you're a speaker who wants to talk Sports Geek, you'd better demonstrate an interest in the sports industry, right?

The Influence Generator

To sell more, or to achieve the goals you've selected, you must map your goals back to topics and your topics back to people who are authoritative on these topics.

People connect with your "WHY" (outside ring), then want to understand HOW (middle ring), and only then care about the WHAT (the product/service you sell).

Our personal brands allow us to show this through the content that we create, which of course always ties back to our "WHY". At Outrageous Digital, our "WHY" is to educate clients which carries through in all of what we do. Once you've determined yours, create a 3-minute video about it.

1a. Map your Personal Branding

Your personal brand is built upon the brands of your ambassadors. Better to leverage the authority that your ambassadors have in the topics you have chosen, as opposed to having to build it from scratch yourself.

Comment on blogs and build your network

As you read through the articles and blog posts that Google delivers to you, post short, insightful comments. Don't just compliment the author; say something interesting and unique enough that is clearly written in response to their article.

Most blog leaders have their own Facebook page for their fans (if you don't already, make one). Save the Facebook friend invite for when you have a bit of a rapport with them as you don't want to seem clingy, desperate, or obsessive.

After you have a few dozen posts where they've responded to your comments, then you may try adding them on Facebook and talking to them personally. Explain your passions and mention that you're also writing about the same topics that they are.

Write your article draft

Make sure you have at least 200 words and take screenshots that show what you're trying to convey. Placing a red box around important numbers is the best way to draw a reader's attention.

Write succinctly. Break up long paragraphs for easier reading. Be clear and concise so that your style shows through. Use the process of MAA (Metrics, Analysis, Action).

Start with the numbers/metrics, give it context by telling the story behind them, and then make recommendations that form a strategy as to what they can do to improve.

Link to thought-provoking content with relevant anchor text

When researching your topic, keep a list of articles you can sprinkle into your final work. These are usually links to blogs you've had contact with or contribute to, so consider how well you're doing with Step 4.

When linking, make sure to avoid using "Click Here!" anchor text. Instead, use text that portrays what the link is about, like ***why you shouldn't steal your competitor's traffic.***

Don't over-promote yourself; otherwise, you'll be labeled as a spammer. It's fine to have a URL in every paragraph or subtopic to drive your point home, but absolutely NO affiliate links or blatant for-profit material.

Give your article a snappy title

Once your article is done, it's time to give it a killer headline that summarizes what your article is about and why your audience should care. Be careful with sensationalized headlines / "viral" tricks. "Wow, I can't believe this actually worked!" is obnoxious and says nothing.

Write down a few headlines that best summarize your article. Ask your friends which one of them grabs their attention the most.

Work out how often you can generate content and how often your blog network can accept it

This is the most important part and the hardest to get down. Whereas, up until this point, you were writing and posting regularly on your own blog, you'll now distribute across many blogs at the same pace.

By now, topics should be plentiful and writing articles should come easily to you. Once you can react fast enough, build a small backlog of articles that you can distribute elsewhere so that you're not pressed for time.

You should contribute to each site in your network at least once a week. Some will accept more; others, less. But most people are happy for free content. Just make sure to consistently deliver high-quality content.

Scaling the writing and editorial process with a team

Once you have a larger network of publications to provide content to, you'll need a team to effectively scale the process and assist with creating content.

Fancy Hands Upwork™ fiverr® WriterAccess

Once you've assembled your content experts, you should assign roles based on the stage of the article. These roles include:

- ✓ **Senior Editor:** Approves the final draft for publication and verifies all facts.
- ✓ **Editor:** Proofreads all drafts and provides additional media/links where applicable.
- ✓ **Author:** Assembles drafts on given topics and gathers initial media to support the article.
- ✓ **Videographer:** Records high quality video/audio of interviews or events on location, too.
- ✓ **Video Editor:** Takes raw videos of interviews or other relevant media and edits them for publication.
- ✓ **Transcriptionist:** Watches/listens to videos and translates audio into text.
- ✓ **Project Manager:** Ensures articles are moving forward and, if the article is for a third party, handles communication and media acquisition.
- ✓ **Amplifier:** Takes finished articles and posts them on social channels, running micro-targeted ads to increase interest.

If you don't have the resources to fill all of these roles, delegate to virtual assistants (VAs) like Fancy Hands, Fiverr, Upwork, or WriterAccess.

Fancy Hands Upwork™ fiverr® WriterAccess

- ✓ Import your customers and leads' emails into Facebook, LinkedIn, Twitter, and Google as custom audiences.
- ✓ Build targets on Facebook and Twitter - direct interests, closest competitors, common interests your customers share, industry working in the media.
- ✓ Create 1% lookalike audience for each major landing page, thank you page, and email list.
- ✓ Amplify a video and create video remarketing audiences.
- ✓ Create 1-day, 30-day, and 180-day audiences (Website Custom Audiences), site-wide, for each major landing page and thank you page.
- ✓ Build Bowtie Funnel Sequences.

☐ Fundamental ☐ **Advanced**

How to take advantage of microtargeting on Facebook — at a crazy cheap price

Facebook ads are powerful. One of our assistants was trying to get my attention. He did so by creating a Facebook ad targeting anyone who lived in Michigan between 30 and 40 years old and worked at BlitzMetrics. Of the millions of users on Facebook, only 80 people met that criteria.

It cost him only 6 cents to do it. And for that price, he was able to bombard BlitzMetrics people with ads. The cost of that inventory was a 30 cent CPM, which means it cost 30 cents to show one thousand ads. So he was able to send 200 highly targeted messages.

Sounds less like advertising and more like super-targeted email marketing, doesn't it?

And, in fact, it is, except for this:

- ✓ You can send these messages without needing someone's email address.
- ✓ You pay only when someone clicks it (yes, its cost per click advertising).
- ✓ An impression is guaranteed when the person opens Facebook (whereas in email marketing, you can only hope that someone will open it).

A move to quality targeting over mass media blasts

Now, do you see how this works? It's quality over quantity. Think about who you want to target as precisely as possible. Where do they work? Where do they live? What kind of car do they drive? What TV shows do they watch? What industry conferences do they attend?

Can't afford $15,000 to exhibit at your favorite conference, plus the $3,000 ship the booth out, the cost of the people to have to man the booth during Expo Hall hours, the promotional material you have to give out, and so forth?

Here's the solution. Run an ad for the three weeks leading up to the conference by targeting fans of the conference.

Need son PR help but you can't afford a PR agency for $10,000 a month? Then let Facebook do the work for you. Run ads that target journalists who write for the The Wall Street Journal, New York Times, Mashable, Forrester, VentureBeat, or whoever.

Just run Facebook ads targeting the competitors of your existing customers.

Let's say that Marriott is your client and you have a great case study with them. Run ads targeting the executives who work at Hilton, Starwood, Motel 6, or whoever. You can bet they want to know what their competitors are doing.

Now imagine that you're a software company that is building relationships with other agencies that resell your social analytics software. The founders of a data visualization agency come to visit you and you'd like to strengthen that bond.

Maybe you spend $1 a day on a microtargeted campaign like the one above, but slice it up to put the ad image more compactly next to the stats.

You absolutely bombard anyone who works at that firm with your message almost 3,000 times. If they have 50 people, that's 60 ads per person. Who cares that we got only 9 clicks (of which 4 happened to become fans)? The goal is not the click, but the awareness.

Total cost: $5.67 in Facebook ads.

While each of these examples might be clever or interesting, the question becomes: How do you scale this?

Ultimately, social media success is about pinpoint precision targets because we're simulating the one-on-one conversations that friends have among themselves. If you want to have 1,000 conversations you need 1,000 different ads and 1,000 different landing pages. Who has the infrastructure, staff or budget to do that? This is were smart automation comes in.

Software companies are good at building code based on a predefined set of rules that can be repeated. Success for your company can't be solved by either a pure agency or a pure software company. The agency can't throw enough people at the problem and the software company can't offer a one-size-fits-all solution to everyone.

Only you can work the magic at your company. As much as we'd like to sell you our software, vendors like us can only assist you in coming up with the creative strategy that resonates best with your customers, the PR strategy that gets the press talking about you, and a unique way to position how you solve your client's pain.

Endorsement is a key part of the marketing process. It shows personal approval and proven application rather than simply saying "It's awesome because we say it is."

You should have a process for collecting user reviews and accolades of your product. Then you can use Facebook's carousel ads to their experiences.

Make sure that these pieces of content that you feature meet certain criteria, such as their expectations, their usage of product, experience level, etc. You'll also need to have a process in place to acquire user consent before featuring them-- not through a heavy-handed disclosure form, but a light compliment/question that requires as little effort as possible on the user's side. like how Nautica shoes do with #GoNautica.

Targeting Planning Checklist

Name 3 direct interests related to your brand.

Interest 1 ______________________________

Interest 2 ______________________________

Interest 3 ______________________________

Name your closest competitors.

Closest competitors are competitors who your customers regard as a substitute - not necessarily the biggest competitors in your industry.

Competitor 1 ______________________________

Competitor 2 ______________________________

Competitor 3 ______________________________

Name 3 interests that your targets share.

Interest 1 ______________________________

Interest 2 ______________________________

Interest 3 ______________________________

Import your customers' and leads' emails into Facebook and Google as custom audiences.

Target		Audience size
List your competitors *Closest competitors are those who your customers regard as a substitute – not necessarily the biggest competitors in your industry.*	1.	
	2.	
	3.	
List your customers' common interests	1.	
	2.	
	3.	
	4.	
	5.	
List 5 workplace based targets (think of influencers in your field)	1.	
	2.	
	3.	
	4.	
	5.	
Create saved audiences for each of these 3 groups, including 1 micro-targeted audience for workplace targets	1.	
	2.	
	3.	

Example Influencer Audiences

Audience Details

View how this audience is defined

Audience Name Sports Media Workplaces - 13k

Saved Audience Details

Location: United States

Age: 18 - 65+

People Who Match: Employers: Bleacherreport.com, Athletics Weekly, ESPN, FanSided, ESPN The Magazine, Sporting News, TheFostGame, Yahoo Sports Fantasy, ESPN Magazine, FOX Sports, TNT Drama, HoopsHype, SB Nation NBA, Sports Illustrated, SportsCenter, Yardbarker, NBC Sports, Basketball Insiders, SportsPro, Sports Center, Professional Sports Publications, RealGM, Bleacher Report Radio, Yahoo! Sports Radio, CBS Sports, NBADraft.net, SLAM Magazine, Yahoo Sports, SB Nation, FOX Sports Arizona, RantSports, Sports Weekly Magazine, Scout Media, RealGM.com, Deadspin, FOX Sports Ohio, Bleacher Report, SportsCenter Inc, ESPN3 or Sports journalism

Edit

Audience Details

View how this audience is defined

Audience Name Media Workplaces - 97k

Saved Audience Details

Location: United States

Age: 18 - 65+

People Who Match: Employers: The Guardian, Ellen DeGeneres, People magazine, SFGate, Time Warner Cable, CNN Sports Illustrated, ESPN, ESPN First Take, FOX Sports, ESPN on ABC, Live Nation Entertainment, TIME, The Late Show with Stephen Colbert, Chron.com from the Houston Chronicle, NBC Nightly News with Lester Holt, Webclinic, Today Show, CBS News, The San Diego Union-Tribune, Live with Regis and Kelly, USA TODAY, iHeartMedia Chicago, Mike and Mike, Detroit Free Press, New York Post, Viacom, Weekly World News, SportsCenter, The Tonight Show with Jay Leno, Good Morning America, USA TODAY Sports, The Dr. Oz Show - Zoco Productions, NBC News, NBCUniversal, NBC Sports, Oprah Winfrey Show, The Late Late Show with James Corden, Fox News, Iheart Media, Steve Harvey Morning Show, New York Daily News Sports, CBS Radio, Jimmy Kimmel, The Buffalo News, Dr Oz Show, Business Insider, Mashable, CNN International, Newsweek, The Dallas Morning News, Newsday, The Sacramento Bee, PBS, FOX10 News, Sports Center, New York Times, New York Daily News, MSNBC, iHeartMedia Inc., CBS Sports, The Star-Ledger, IHeartMedia, iHeartMedia Portland, Miami Herald, XFINITY, The Seattle Times, Dr. Mehmet Oz, Jimmy Kimmel Show, The Charlotte Observer, CNN Sport, Turner, ESPN3, Star Tribune, Jimmy Kimmel Live!, Hartford Courant, The Chicago Sun-Times, Late Show With David Letterman, Live With Kelly And Micheal, azcentral, The New York Times, The Boston Globe, CNN, Orlando Sentinel, Las Vegas Review-Journal, Washington Post, Chicago Tribune, Los Angeles Times, Omaha World-Herald, Sun Sentinel, Orange County Register, CNET, ISM Sports, CBS Sports Network, ABC News, The Columbus Dispatch, The Kansas City Star, O, The Oprah Magazine, NBC, Tampa Bay Times or The Virginian-Pilot

Edit

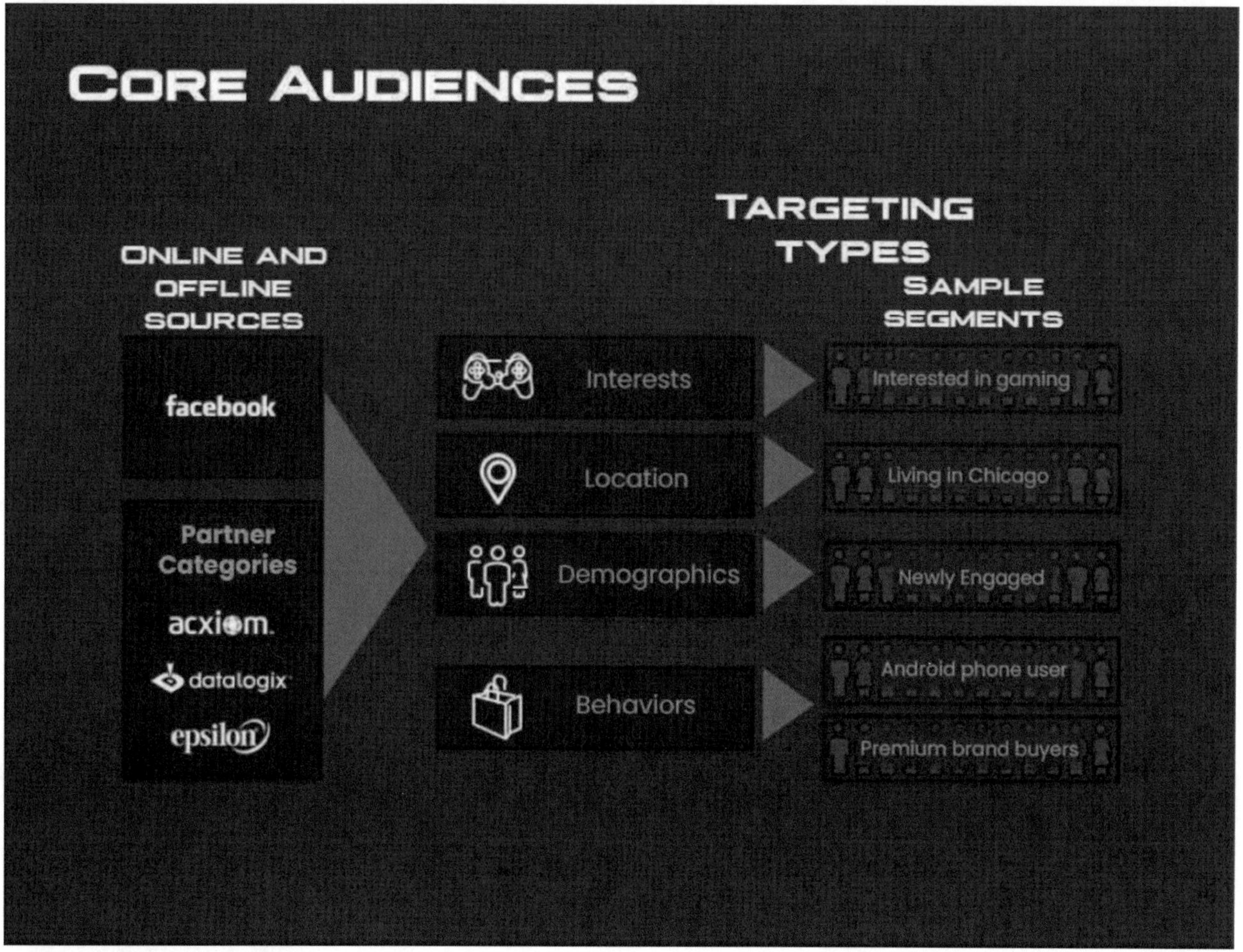
CORE AUDIENCES
ONLINE AND OFFLINE SOURCES
facebook
Partner Categories
acxiom.
datalogix
epsilon
TARGETING TYPES
Interests
Location
Demographics
Behaviors
SAMPLE SEGMENTS
Interested in gaming
Living in Chicago
Newly Engaged
Android phone user
Premium brand buyers

THREE TYPES OF AUDIENCES

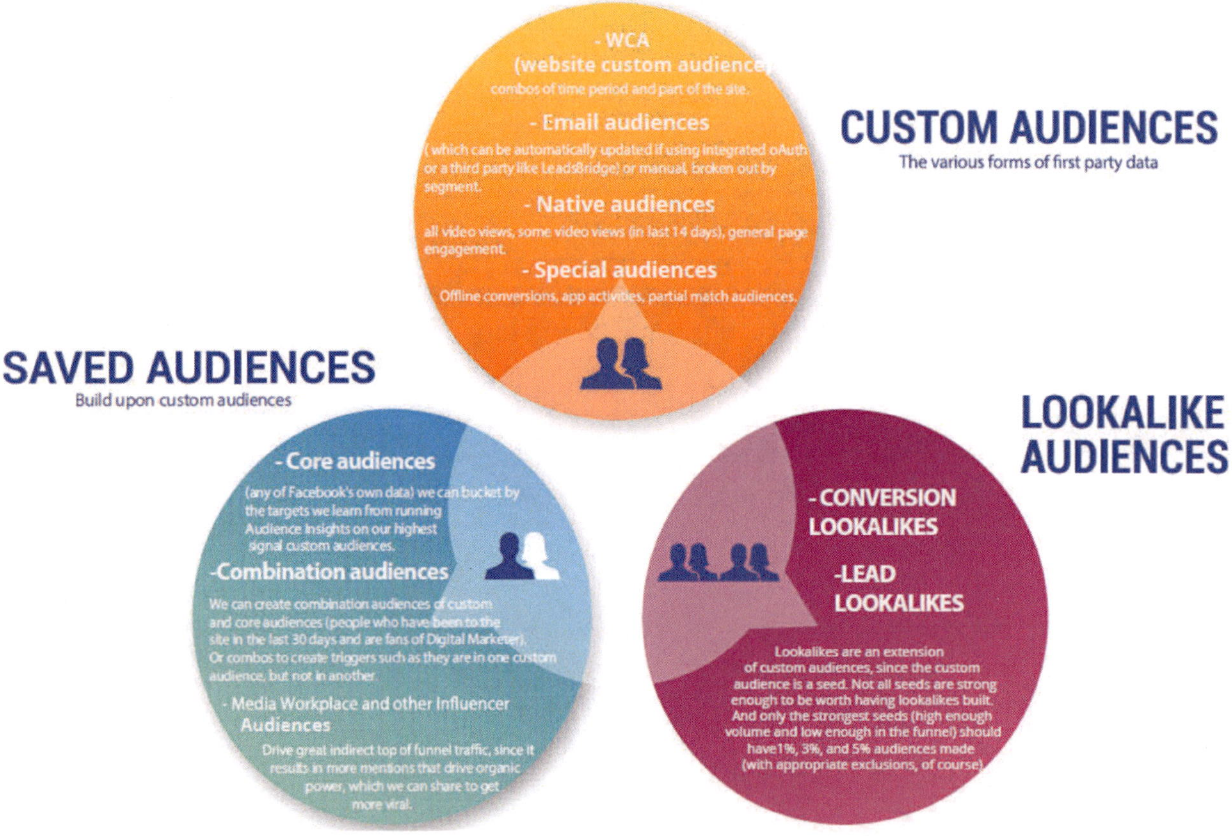

Targeting the right audience is based on the steps of the funnel. The Warriors have a high number of Fans (11.3MM in Sept 2018) and a large website remarketing audience, both from ticket sales and the Warrior Team Store.

Remarketing means to show proven content to an audience which is more likely to convert than the average Facebook fan.

Therefore, the funnel is designed to increase remarketing audiences and nudge people to buy tickets and merchandise.

Amplification

- ✓ Boost top 3-5 Facebook Posts to at least 1 Saved Audience per persona.
- ✓ Boost optimization: 4 Stages.
- ✓ Set up remarketing ads for 1-day landing page abandoner's on Google Ads and Facebook.
- ✓ Create media inception ads using the Dollar A Day strategy.
- ✓ For each unpublished post, use tracking (UTM) parameters in the URL.
- ✓ Create unpublished posts, boosted for website clicks and conversions.

☐ Fundamental ☐ **Advanced**

How To Set Up A Facebook Ad Campaign In 60 Minutes — And Look Good

We'll do three campaigns:

1. News Feed Exposure

This ad is to make sure that your fans see your posts. If you're a big brand, only a few percent of fans may be seeing your posts in the News Feed, versus 10 percent to 20 percent for others. Select your page from the drop-down, then the middle option for "promote page posts," and then the check-box for promoting the most recent post.

Facebook tries to create another ad to accompany it — keep it there. It's a page post like a story to amplify activity that occurs on your post.

Check the option to target only people connected to your page. If you have under 200 fans, this campaign won't do much for you, as not many fans will see your messages. You'll need page post ads targeted at "anyone," but with a bunch of precise interest targets, which we won't cover here.

Name this campaign "page-post-ads" and set a budget equal to your fan base divided by 100, but at least a dollar. So if you have 500 fans, budget $5. If you're not in the U.S., U.K., or Canada, do one-half of this,since traffic is cheaper.

2. Job Titles (The Business-To-Business Secret)

If you're B2B, you need to hit people who work at certain companies in certain departments. In the precise interests box, enter these job titles. Facebook will keep suggesting more, so keep adding the relevant ones. Make sure the number is at least 40, but perhaps not over 10,000.

But even if you're not B2B, job title targeting will still work for you. If you sell children's books, target teachers in your neighborhood — the superintendents, perhaps. If you're a cosmetic surgeon, target other medical specialists who'd drive referrals to you. If you're a wealth advisor, put in "vice president" and "CEO" to reach rich people.

Put $1 per day on this for every 200 people you are targeting. If you're a local business, this audience shouldn't be over a few hundred.

Create a new ad choosing "get more likes." Then enter a headline and text. Uncheck the box that says "only people not connected to … ," since we want fans to see this, too. Facebook creates a tag-along ad called a "page like story," which you should keep.

Your budget should be whatever you're comfortable spending to grow your fan base. Expect your cost per fan to be between 20 cents and $2, depending on your industry and how clever your copy is. As a general rule of thumb, your budget should not be more than one penny per user, per day, or else you risk burning out the ad, spamming users, and wasting your money.

3. Interest Targets

Hit “create a similar ad” below the ad you made earlier, and remove the job title targets from the precise interests box. Add in targets that are literal, lateral, and competitive.

Competition is easy: The companies you compete against, you’re hitting their fans. Choose partners, too, if you’re B2B.

Literal: If you sell Ford Mustang parts, find keywords like "working on my

Mustang” or “Ford Mustang.” If you’re a service business, your Facebook ads might not convert, since this isn’t Google, where people are searching right then for something. If you’re selling something considered, such as a mortgage, car, or vacation package, try broad category targeting (just below the box with precise), where you can target people who just moved, are looking to buy a car, have kids, are having a birthday that day, are Hispanic, have a Samsung phone, etc.

Lateral: This is where the targeting fun really is. If you’re Jack Daniel’s, you might target certain country bands. If you’re a marijuana dispensary, target folks in Portland who listen to Bob Marley. You get the idea.

Hit “create similar ad,” each time, giving each ad a descriptive name. It takes only 30 seconds to make an ad, so you might as well do a dozen or two.

If you lump all of your targets in one ad, you won't know how each of these interests are performing. If the combined audience size of an ad is under a few hundred, then there's not enough data to warrant creating another ad.

Dollar a Day

1. Determine the audiences you what to influence.
2. Find high authority mentions you would like to amplify.
3. Run Facebook ads at $1 a day for each high authority link.
4. Reply to everyone who engages in your amplification efforts.
5. Collect and identify authority mentions.
6. Continue to build momentum in your PR machine by cycling in new content and audiences.

Amplification Overview

Specialist Setup

Step 1-3 takes 2 minute

- ✓ Open a new tab and log in with your personal Facebook account at *business.facebook.com* and go to the client's Ad Manager.
- ✓ Open your Campaign Segmentation Plan.

Create Campaign

Step 1-3 takes 2 minute

- ✓ Click Create Campaign.
- ✓ Enter Campaign Name. Choose Buying Type. Choose Objective. Choose "Auction" for Buying Type unless otherwise stated.
- ✓ Click Create.

Create Ad Set

Steps 1-3 take 2 minutes

- ✓ Click on the Ad Sets icon.
- ✓ Click Create Ad Set.
- ✓ Choose an existing campaign.
- ✓ Enter Ad Set name according to the naming conventions.
- ✓ Click Create.
- ✓ Select the budget type into a empty field.
- ✓ Choose Start Date. If you want to put an end date, choose end run on. Else, choose don't schedule end date, run as ongoing.
- ✓ Click **Audience.**
- ✓ Fill out the Custom Audience field (if any).
- ✓ Fill out the Location field (compulsory).
- ✓ Fill out the Age field (compulsory).
- ✓ Fill out the Detailed Targeting field (if any). Include any AND/OR logic if indicated.

- ✓ Fill out the Connections field (if any)
- ✓ Click **Save.**
- ✓ Choose placement
- ✓ Choose Optimization and Bidding. Use the default option unless otherwise stated.
- ✓ Choose Delivery Type. Use the default option unless otherwise stated.

Create Ad

Step 1-18 takes 20 minutes

- ✓ Click Ads icon.
- ✓ Click **Create Ad.**
- ✓ Choose an existing campaign and ad set.
- ✓ Enter Ad name according to the naming conventions.
- ✓ Click **Create.**
- ✓ Select client's Facebook Page and Instagram Account (if applicable).
- ✓ If promoting a post that the client has published on their Facebook Page or Instagram Account, click Use Existing Post. Else, proceed to Step 9.
- ✓ Obtain Post ID and insert in the empty field. Proceed to Step 10.
- ✓ To create a single image ad, tick the box beside Ad with an image or video.
- ✓ To create a carousel ad, tick the box beside Ad with multiple images or videos in a carousel (Recommended).
- ✓ Fill out the Headline field (compulsory).
- ✓ Fill out the News Feed Link Description field (if applicable).
- ✓ To use an image, click **Select Image.**

- ✓ On the popup, find and upload the image.
- ✓ Click Confirm.
- ✓ To use a vieo, click on the radio button beside Video.
- ✓ Click Select Video
- ✓ On the popup, find and upload the video.
- ✓ Click Select.
- ✓ Choose the Call To Action from the drop-down menu.
- ✓ Repeat Steps 2 to 5 (including the subsets) to create another ad

Upload Ads

Step 1-18 takes 20 minutes

- ✓ Check that you have implemented the Checklists 02 to 04 correctly.
- ✓ Review Changes in the top-right corner.

Targeting

Location

Target by location when relevant. If you're promoting an event in Boston, set your location within a 50-mile radius of the city. Otherwise, don't target outside of the main industrialized countries unless that's where your market is.

oCPM

Facebook optimizes ads for parameters such as age, gender, and demographic, so there's no need to specify these targets unless, of course, your market is .g. narrowly defined (e.g.34-year-old females from London).

Workplace/Interest Targeting

With specific workplace and interest targeting, narrow an audience down to influences in industry that can do PR for you. If you want to be featured in CNN, boost content to people who list CNN as their employer, job title "Journalist", and so on. All it takes is a dollar a day to make ripples by "incepting the inceptor's".

Boosting

"Fuel On The Fire" and Saved Audiences

Experiment by boosting posts to many different audiences at the same time and monitor for "winners" that engage well. . That is, identify what's working (creative, title, targeting, etc.) and amplify it—spend way more money on it.

Add successful interest and workplace targets to saved audiences in Ads Manager, which you can later access from the News Feed.

Now all you need to do is publish, select a saved audience that aligns with the content, and hit boost.

- ✓ Apply Metrics Decomposition.
- ✓ Compare current period against last period.
- ✓ Using Audience Insights, create new saved audiences.
- ✓ Review budget allocation by channel and ad set based on performance (watch for statistical noise).
- ✓ Increase relevance positive and negative audiences.
- ✓ List 3 to 5 top recommendations to execute in the next 7 days.
- ✓ Apply Top N to the data set and explain the results in terms of Goals, Content, and Targeting for each level.
- ✓ Refine lookalike audiences.
- ✓ Update Success Tracker.

☐ Fundamental ☐ **Advanced**

Beginning Optimization

The more complex your campaigns, the more time it takes to optimize them. The more ads you've created, the more "weeding" your garden needs. So don't go crazy making a ton of ads at once - just 5-10 per campaign within the 3 campaigns.

That's why we advocate the simple three-layer campaign strategy of Awareness, Consideration, and Conversion. These campaigns are "always automatically.

When you have only a handful of ads in a campaign, you can quickly see whichones to kill off and which ones to create more variations from. This is a modified version of "winner stays on," where you keep cloning the top performers with similar audiences, but slightly different.

Remember to create ad groups of one or multible interest filters. Most of the ads have 3-4 filters each. Some have 7-8 filters on them. Having multiple combinations is calles "onion targeting".

For a new set of campaigns, you should check in 2-3 times a week, spending only 15 minutes each time. Better to do a few adjustments over time than try to create a ton of ads at once and have only one cycle of optimization.

You want to work with just a few ads per campaign, allowing you to nimbly pause a couple ads (don't hit "delete") and create a couple new ones.

This allows you to quickly get to statistical significance-a minimum number of clicks to be able to tell whether the difference in performance between a group of ads is a real difference or random noise. A general rule of thumb fo people who forgot what they learned in stats class-get 10,000 impressions or at least 20 clicks for any ad. Any less and it's noise-you can't really tell.

Once your pilot is successful

Once you reach a point where you've been able to prove your campaigns are successful, then you can scale up the daily budgets. That also means you can

readjust your audience size to reflect the larger budgets.

At this point, you'll be creating more and more audience targeting combination to include custom audiences and FBX.

You then clone your three campaigns so you now have two of each-- two Awareness, two Consideration, and two Conversion. Except you have one "test" campaign and one "production" campaign for each type of campaign.

The test campaign now becomes what you've been experimenting with, while the production campaign is much larger. When you find a winner in the test campaign, you copy it over to the production campaign.

You wouldn't test out new ads in the production budget since it could risk eating up the larger budget on a larger audience. For those who are veterans of Google Ads, this is called the "paste and stick" method.

Once your pilot is successful

Once you reach a point where you've been able to prove your campaigns are successful, then you can scale up the daily budgets. That also means you can

readjust your audience size to reflect the larger budgets.

At this point, you'll be creating more and more audience targeting combination to include custom audiences and FBX.

You then clone your three campaigns so you now have two of each-- two Awareness, two Consideration, and two Conversion. Except you have one "test" campaign and one "production" campaign for each type of campaign.

The test campaign now becomes what you've been experimenting with, while the production campaign is much larger. When you find a winner in the test campaign, you copy it over to the production campaign.

You wouldn't test out new ads in the production budget since it could risk eating up the larger budget on a larger audience. For those who are veterans of Google Ads, this is called the "paste and stick" method.

Super pro tips

Keep in mind that if the audiences are small for an ad, it would cap out on the audience well before it capped out on the daily campaign budget. Budgets are set at the campaign level, not the ad level, so you still need to be careful.

Facebook doesn't have frequency capping or ad rotation features natively in Power Editor, the regular ads interface, or the API - but we hear it's coming. Meanwhile, that means you have to watch your frequency per ad carefully, as well as look at Frequency at the campaign and account level.

If you're doing a great job at onion targeting, some of your users are seeing News Feed posts, dark posts, and sponsored stories.

You know that reach x frequency = impressions. Facebook shows reach and frequency, but not impressions, but you can calculate it easily. They choose to show reach and frequency instead because it's more actionable. Your reach tells you who are exposed to your ad, while frequency is how often those folks see it.

If you have a reach of 10,000 people and an average frequency of 10, then you've served 100,000 impressions. A synonym for reach is unique impressions - or unique users who have seen at least one impression.

In general, the News Feed placements (desktop News Feed and mobile News Feed) are better for awareness and consideration campaigns. For conversion campaigns, sometimes the RHS placement performs better - better CPC or better CPA.

You just have to test. It may be that non-social businesses (think of things like hemorrhoid cream, industrial plastic extrusion machinery, or personal injury attorneys) produce content that people wouldn't want to actually react to, even if they're interested.

We optimize using our Facebook Optimization Checklist but with specific adaptions for the Warriors:

Make sure campaigns are turned on/off depending on the games they are promoting

Find the best performing page posts and boost them before, during, and after each game.

- ✓ Find the best performing page posts and boost them before, during, and after each game.
- ✓ Monitor the mix between remarketing, lookalike and interest-based ad sets.
- ✓ Use our Top N and MAA approach to optimize ads.
- ✓ Focus on campaigns, ad sets, and ads with the biggest budget and conversions.
- ✓ Look at metrics (CTR, CPM, Conversion Rate).
- ✓ Create an analysis (Why is CTR high/low?).
- ✓ Take action (Change the ad, ad set, or targeting options.

Boost Optimization

Kill 90%

If after 7 days of a dollar a day ($7 spent total), we don't see results (meaning the cost per engagement or average watch time isn't good), then just let it die. Do this 90% of the time-- don't waste your money.

Expand 5%

If it's pretty good, we can then add another $30 for 30 days. Perhaps 5% of the time you'll find a worthy post-- usually a video that gets above 20 seconds in average watch time or a regular post with over 10% engagement (interactions divided by impressions).

Switch-boost 3%

If it's a winner, change the audience we're targeting in the same boost, right from the timeline or Pages Manager. Usually, we'll initially boost to generate high traffic (BuzzFeed audience to a post on BuzzFeed), switching to a media workplace audience a few days later. Or we'll switch to whatever audience we currently want to incept-- perhaps we're speaking at the Social Media Marketing World conference, so we'd switch a bunch of posts to that audience temporarily. Some posts live forever-- they keep producing for months or even years, often triggered by custom audiences.

Stack-boost 2%

If it's a "unicorn" (off the charts performance), duplicate the ad set inside Ads Manager so that we have another ad set running concurrently. When something is doing really well, we want to have multiple audiences seeing a post, which means we need multiple ad sets. This is not possible inside the timeline, but only takes a minute inside Ads Manager or Power Editor. The risk of a unicorn is that if you run it too hard, it will burn out. We've seen a few companies accidentally kill their unicorns by literally spending millions on them, but not having any backups. Create "unicorn children", as Larry Kim likes to call them-- derivative pieces of content so you can get a similar effect from similar posts.

4 Phase Boost Sequence

SEED
ORGANIC
The initial post where you share your story--something of emotional meaning, ideally within a one-minute video. Could potentially be an article you've written or a mention. Pull these from the outside ring of your Topic Wheel. Evaluate success by benchmarks for post type (video, image, lead ad, etc.).

BOOST
PAID
If you're doing well organically , put a "Dollar a Day" behind it to amplify what the algorithm says is already doing well. Choose an audience that would have the highest engagement rate, so your "spill-over" into additional organic reach/engagement is high. Might be a workplace audience, too. Follow the 4 things to do after a boost to keep tuning.

SHARE
ORGANIC
Provided others are picking up your content, then share their post, saying thank you and tagging them. Use the free monitoring tools listed in the "Influence Generator Course" (CrowdTangle, mention.com, Google Alerts, Twittter alerts, FB notifications) to capture these mentions.

VIRAL
PAID
When your share has accumulated more than 10,000 views or 1,000 likes, boost for $1 a day or more against large interest audiences. If video, then your shared post will continue to accumulate views in all places that video is shared. Flag this post in the "Greatest Hits" section of your Content Library, so it can run forever.

Reporting is the logical conclusion of the MAA process. We are setting up the following reports

- ✓ Weekly Reports via Success Template
- ✓ Sponsorship infographic
- ✓ Digital Overview Infographic

The weekly report shows

- ✓ Delivered Impressions
- ✓ Delivered Clicks
- ✓ %CTR
- ✓ Post Click Conversions
- ✓ Post Click Revenue
- ✓ Post View Conversions (28 days) Post View Revenue
- ✓ Tickets Sold
- ✓ Conversion Value (if available - depending on Ticketmaster
- ✓ Total Conversions Combined
- ✓ Total Revenue Combined
- ✓ Spend
- ✓ Cost/Click
- ✓ Cost/Conversion
  %ROI

The Sponsorship Infographic

This infographic relies on our online marketing performance and is used to convince potential sponsors to partner with the Golden State Warriors. The following screenshot shows an example of how one page in a sponsorship infographic could look like (template):

NAMING CONVENTION FOR ADS

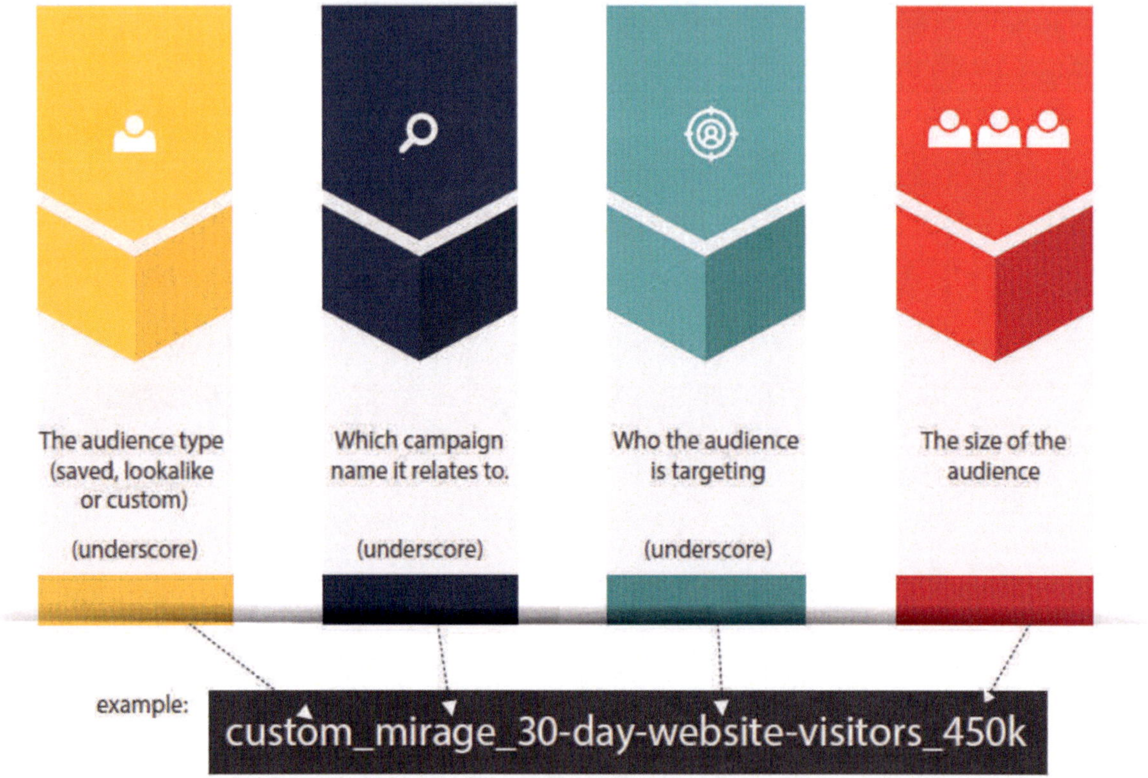

By keeping up-to-date records, it allows us to track fluctuations in audience sizes. Plus, we can plan and build out campaigns & creatives where we can target specific audiences.

Join my Podcast, **SOCIAL MEDIA ENGINE** as we discuss how to get the most from this book!

Every Tuesday at 2:00 PM EST

SPECIAL OFFERING

Digital Plumbing report for **$995.00**

Billed through

VENDASTA

Outrageous Digital provides online marketing solutions, for small to medium businesses. Our unique solutions include, **do it yourself (DIY), do it for me (DIFM) and do it with me (DIWM) services.** This makes it possible for small and medium-sized businesses to reduce client acquisition cost, and prove a positive return on investment in today's marketplace.

Outrageous Digital

100 W Big Beaver Rd Troy, MI 48084

www.outrageousdigital.com
steven@outrageousdigital.com

blitzmetrics

blitzmetrics

PARTNER

THIS DOCUMENT CERTIFIES THAT

Steven Keltsch

CEO | Outrageous Digital

Having successfully completed all the requirements has been granted with

CERTIFIED DIGITAL MARKETING AGENCY

certificate and has been awarded all the privileges pertaining to this credential.

This certificate can be verified at: https://bit.ly/Sh14683842

09/10/2020
Date

IN PARTNERSHIP WITH
VENDASTA

Dennis Yu
CEO of BlitzMetrics

Thank you

Made in United States
Orlando, FL
04 November 2026

90201572R00064